METEORS

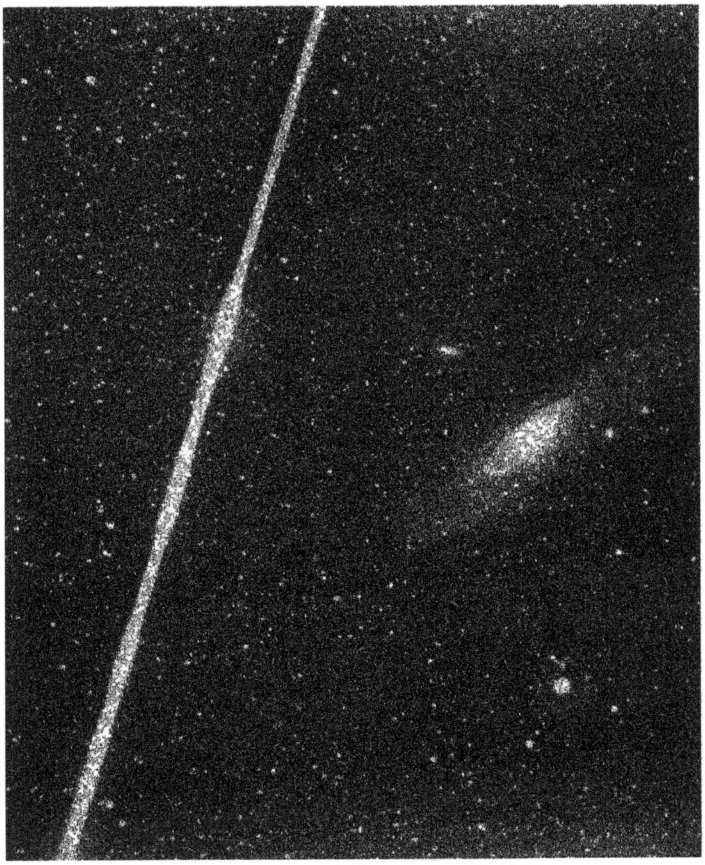

THE GREAT BOLIDE OF SEPTEMBER 12, 1923

Photographed by Josef Klepesta, at 21^h 55^m 25^s G.M T , at Prague Observatory

METEORS

BY

CHARLES P. OLIVIER
ASSOCIATE PROFESSOR OF ASTRONOMY
AT THE
UNIVERSITY OF VIRGINIA
AND
ASTRONOMER
AT THE
LEANDER McCORMICK OBSERVATORY

BALTIMORE
WILLIAMS & WILKINS COMPANY
1925

Made in United States of America

Published March, 1925

COMPOSED AND PRINTED AT THE
WAVERLY PRESS
By the Williams & Wilkins Company
Baltimore, Md., U. S. A.

TO

THE MEMORY OF MY FATHER

GEORGE WYTHE OLIVIER

THIS BOOK IS DEDICATED

PREFACE

One of the few departments of astronomy that has not been brought up to date in a recent book is that dealing with meteors. An excellent book on meteorites, by O. C. Farrington, was published in 1915, and during the past few years several important researches on meteors proper have appeared, but no general work on meteoric astronomy has been published since 1871. Such a book, however, is urgently needed, because much of the most valuable recent work on this subject has appeared in foreign periodicals inaccessible to the average reader. In this book many of the most important of these articles are briefly reviewed, and numerous references are given enabling the reader to follow up the subject, if he so desires. The more mathematical and theoretical sections have been segregated and may be omitted without destroying the continuity of the book as a whole for general reading and reference. Practical examples, given in detail, are added for the use of the amateur who may wish to compute his own orbits.

As Schiaparelli's classical work *Sternschnuppen* is now very difficult to obtain, and has never appeared in either an English or French edition, many sections of it have been adopted. The methods for computing orbits are taken almost unchanged from the publications of Lehmann-Filhès. The method for computing real heights, due to Schaeberle, is reproduced from one of the Lick Observatory publications, with the kind permission of Director W. W. Campbell. Acknowledgment is here made for all the above material.

The writer accepts full responsibility for all opinions and criticisms which appear, unless definite references are made to others. It is only fair to state this fact since his opinions, particularly upon radiants and the interpretation of observations, differ radically in certain respects from those of others who have done important work in meteoric astronomy. Further it is clearly to be understood that the opinions here expressed supersede any formerly published by him, as in more than one phase of the subject longer study and experience have caused him very considerably to modify those previously held. And these in turn are, of course, liable to further change since, in a subject like meteoric astronomy, few opinions can be final.

Most grateful acknowledgment is due to the Superintendent of the United States Naval Observatory, Captain E. T. Pollock, for kindly placing at the writer's disposal the facilities of the splendid library of that institution. The preparation of this book would have been impossible without the use of this library, where most of the actual writing was done, while on leave of absence from the University of Virginia. The writer is further under great obligations to the following gentlemen for advice and assistance, especially in reading various parts of the manuscript: Prof. W. S. Eichelberger, Prof. A. Hall, Prof. G. A. Hill, Dr. W. D. Horigan, Dr. W. J. Humphreys, Prof. F. B. Littell, Dr Geo. P. Merrill, Dr. H. R. Morgan, Mr. G. H Peters, Dr. Thomas L. Watson, Mr. C. B. Watts, and Mr. J. D. Wise. Further grateful acknowledgments are due to Prof. Edwin B Frost, Mr L E Jewell, Prof. Josef Klepesta, Dr. W. J. S. Lockyer, Dr Geo P. Merrill, Dr. Thomas L. Watson, The Astrophysical Journal, the Lick Observatory, Mt Wilson Observatory, and Yerkes Observatory for the use of illustrations.

Finally the greatest acknowledgments are due to my wife, Mary Frances Olivier, who has aided me in every way in the preparation, criticism and correction both of the manuscript and of the proofs.

CHAS. P. OLIVIER

April 9, 1924
Washington, D. C.

CONTENTS

CHAPTER I

Historical Introduction

CHAPTER II

How Meteors Are Observed

CHAPTER III

Comets

CHAPTER IV

The Leonids

CHAPTER VIII

THE HALLEY'S COMET METEORS AND THE PONS-WINNECKE'S COMET METEORS

CHAPTER IX

RADIANTS

CHAPTER X

DO STATIONARY RADIANTS EXIST?

CHAPTER XI

SUGGESTED EXPLANATION OF STATIONARY RADIANT

CHAPTER XII

THE APPARENT PATHS OF METEORS WITHIN THE ATMOSPHERE

CHAPTER XIII

METEOR TRAINS

CHAPTER XVII

The Formation of Meteor Streams

CHAPTER XVIII

The Perturbations of Meteor Orbits

CHAPTER XIX

Influences of Meteors upon the Earth

CHAPTER XX

Discussion of the Formation of the Bielid and Perseid Streams

CHAPTER XXI

Meteorites

CHAPTER XXII

Are Lunar Craters Meteoric in Origin?

CHAPTER XXIII

Origin of Meteors, Fireballs and Meteorites

CHAPTER XXIV

Conclusion

LIST OF ILLUSTRATIONS

ABBREVIATIONS AND DEFINITIONS OF TERMS FREQUENTLY USED

Astronomical Unit: Earth's mean distance from sun = 92,900,000 miles.

Unit distance. One astronomical unit.

Unit velocity. Orbital velocity of the earth at unit distance from sun. When no unit is mentioned for a distance or velocity, the above are inferred.

Parabolic velocity: Velocity of a particle, moving in a parabola around the sun, at unit distance from the sun

Parallax of sun. Angular value of earth's radius as seen from the sun at a distance of one unit.

Elements of orbits·

- Semi-major axis = a.
- Eccentricity = e.
- Inclination to plane of ecliptic = ι
- Longitude of ascending node = Ω
- Longitude of perihelion point = π.
- Period = P.
- Mean daily motion = μ.
- Perihelion distance = q.
- Radius vector of orbit = r or R.

A. M. S.· American Meteor Society.

B. A A British Astronomical Association.

R. A. S.: Royal Astronomical Society.

M 1. Abbreviation for: *175 Parabolic Orbits*..., Transactions American Philosophical Society, N S , 22, Part 1

M 2· Abbreviation for· *126 Parabolic Orbits* , Publications of the Leander McCormick Observatory, 2, Part 4.

M 3· Abbreviation for· *349 Parabolic Orbits* , Publications of the Leander McCormick Observatory, 2, Part 7. (*M* 1, *M* 2, and *M* 3 contain the work of the American Meteor Society and the writer's personal work.)

CHAPTER I

Historical Introduction

Nearly everyone has seen star-like objects shoot across the clear, night sky, often leaving behind them for a moment or two phosphorescent streaks or trains. In popular language these objects are called shooting stars, and many people still believe they are stars which, displaced from their former positions in the heavens, "fall" to some sudden end. A smaller number of us, on one or more occasions, fortunately have witnessed the passage of some vastly larger celestial visitor across the heavens—a phenomenon that leaves an impression not easily forgotten by even the most unobservant.

It will surprise many to learn that a study of these short-lived bodies forms an important branch of astronomy. To all classes of these bodies, large and small, the very general term meteor may be applied. It is the purpose of this book to describe these bodies, to give their history, and, so far as possible, explain the laws which govern them and the theories now held as to their origin and present place in the general scheme of the universe.

In civilized countries there remains but little of the former superstitions and fears concerning the appearance of any unusual phenomenon in the skies. A century or two ago the case was very different with the masses, and in medieval and ancient times, as attested by the respect in which astrology then was held, these things seriously affected the minds of great men, as well as of the uneducated. Indeed in ancient times not only were the daily and seasonal changes, brought about by the sun, and the motions of the moon and stars of interest as affecting the people's daily, national and religious life, but the appearing of such uncommon sights as a total eclipse, a great comet, or some brilliant meteor shooting across the sky aroused intense interest born of awe and fear. Hence records of such occurrences go back fully 2500 years or more.

The earliest accounts of the falling of meteors are found in the records of that ancient people, the Jews, who, though highly cultured in so many ways, paid little attention to natural science. This remarkable natural phenomenon was believed by them to be supernatural, and to this alone the preservation of the account is due. We find in the Book of Joshua, 10 : 11, as follows: " . . . . as they fled before Israel and were in the going down to Beth-Horan, that the Lord cast down great stones from heaven upon them unto Azekah, and they died: they were more which died with hailstones than they which the children of Israel slew with the sword." The evidence is overwhelmingly strong that we have here the account of the fall of a number of meteorites rather than hailstones. That the narrator doubtless exaggerated the relative number of men killed by them is wholly immaterial and to be expected under such unheard of circumstances. Indeed that anyone should be killed by "stones from heaven" probably was so wonderful to him that he could not have invented the story had he so desired.

The next reference to meteors is found in the Chinese annals for 687 B.C. It is given by Biot[1] as follows: "(March 23), during the night the fixed stars did not appear, although the night was clear. In the middle of the night, stars (des étoiles) fell like rain." The account is translated in another way by Abel-Remmat[2] who makes the last part read: "there fell a star in the form of rain." It therefore seems that a meteorite was referred to and not a meteoric shower. We incline to this view even though the next reference in Biot, for 644 B.C., speaks specifically of "five stones falling."

Pliny in his *Natural History*, **2**, 5, 8, speaks of a great meteorite that fell at Aegospotamos in Thrace in 467 B.C. This was described as being as large as a cart and being of a burnt color. It was held in veneration by the inhabitants of the country. In all, Greek and Roman writers give us four accounts of falls, while the Chinese record sixteen cases during the same length of time. Nevertheless the Ionian school of thought in Greece early assumed the cosmical origin of meteoric stones.[3] Humboldt mentions several legends which probably had their origin in the fall of meteorites.[4] Among these may be mentioned the Sythian saga of the sacred gold, which fell burning from heaven, and remained in possession of the Golden Horde of the Paralatae (Herod, **4,** 5–7) He also gives a Mongolian

[1] *Catalogue Général des Étoiles Filantes . en Chine*, 1841

[2] *Jour. de Phys* , 1819.

[3] *Cosmos*, **4**, 206.

[4] *Cosmos*, **1**, 115.

tradition according to which a black fragment of rock, 40 feet in height, fell from heaven on a plain near the source of the Great Yellow River in Western China.

There is a sound reason for believing that the worship of meteorites was one of the very first forms of idolatry.[5] This opinion is based upon the direct and indirect evidence we have, in classical literature; that meteorites have been found carefully interred in Indian burial mounds in the United States; that one was found in an Aztec temple; and that in some cases savage, or semi-civilized people, still consider them holy. It is known that the Phrygian stone, which was carried to Rome in 204 B C., had long been worshipped at Pessinus as Cybele "the mother of the Gods." This meteorite is described as a black stone, in the figure of a cone, circular below and ending in an apex above. There is reason to believe that the Palladium of Troy, the sacred shield of Numa at Rome, and the image of Venus at Cyprus, were meteorites.

That a similar origin can be assigned to the image of Diana at Ephesus is proved by the very fascinating account of the riot at Ephesus that followed the preaching of St. Paul. In Acts, 19: 35, is given the speech of the town clerk of that city which contains in part: " . . . . what man is there that knoweth not that the city of the Ephesians is a worshipper of the great goddess Diana and of the image which fell down from Jupiter?" We here have evidence[6] that the image was commonly believed to have "fallen from heaven," and what else could it have been but a meteorite, especially as we know that such objects were held in reverence?

It is well known that the sacred stone built into the northeast corner of the Kaaba at Mecca is a meteorite, whose history goes back of 700 A.D., though Mohammedan prejudice has not permitted the analysis of a fragment.

Of all peoples the Chinese have made the most numerous records of every kind of meteoric phenomena. These accounts were carefully translated and published about 1841 by Edward Biot in a work entitled *Catalogue Général des Étoiles Filantes et des Autres Météores Observés en Chine pendent 24 Siècles.* Among these are numerous

[5] H A. Newton, *Am. Jour. Sci.*, (4), **153**, 1, 1897.

[6] The opinion has also been expressed that it was the base on which the statue rested that was a meteorite. Perhaps this is the more probable explanation.

accounts of the fall of stones to the earth, the most ancient being in 644, 211, 192, 89, 86–81, 38, 29, 22, 19, 12, 9 and 6 B.C. The one in 86–81 (exact year uncertain in text) is of particular interest as it is described as falling "upon the palace of Wang-tsai" which was in what is now called the northern part of Pe-chi-li. In most of the cases not only the exact day of the year and place of fall are recorded, but how many pieces fell. For instance in that of 22 B.C., there were eight stones, the largest number mentioned on any of the above dates. We will have occasion to return to the Chinese records when the question of meteoric showers is discussed. Enough, however, has been said to show that we have excellent historical evidence of the fall of meteoric stones as earlyas 2500 years before the present. It is most unfortunate that all traces of these older meteorites seem to have been lost. Indeed the Chinese do not record in what manner the stones were disposed of, or even whether they were considered valuable.

The oldest meteorite still preserved, whose exact date and place of fall are known, is the one which fell between 11:00 and 12:00 in the morning of November 7, 1492, near Ensisheim, Alsace. It was described as falling in a wheat field and penetrating 5 or 6 feet into the ground Its weight was 260 pounds, and its fall was accompanied by a loud clap of thunder and a prolonged and confused noise, heard at a great distance. The Emperor Maximilian, who happened to be near, had two pieces broken off, and the rest of the stone hung in the church at Ensisheim. The learned men who saw it all appear to have regarded it as a miracle, as they are reported to have said they had never heard of anything like it. It is most unfortunate that other rulers did not take equal care to preserve the meteorites which must have fallen throughout Europe during those times

We have seen that the Greek Ionian school believed in the cosmical origin of meteorites and that the Greeks and Romans were well aware of the fact that such bodies fell from outer space upon our earth. During the middle ages various records, more or less accurate, were made of meteorites, but as natural science began to take on new life, with the general revival of learning, the opinion gradually gained ground that stones could not fall from heaven, *hence* they did not. Therefore during the eighteenth century many scientists ridiculed the credulity and superstitions of those persons who claimed to see such falls, and of the rest of the people who believed in "what was physically impossible."

The Great Meteor of February 21, 1922

Photographed by J. Bosler and V. Nechvile, about 21^{h} 40^{m}, at Paris Observatory

We now pass on to the latter part of the eighteenth century. A number of meteorites had fallen during the preceding centuries, and many vivid accounts had been given of the phenomena by eyewitnesses. Yet, in the face of all this evidence, we have an example of stupidity and bigotry, exhibited by the foremost body of scientists of the day—men who doubtless considered themselves, and were so considered by others, the most advanced and "modern" of their time—which for all ages should stand as a warning to any man who feels that he can give a final verdict upon a matter outside his immediate experience. The Académie des Sciénces, having sent a commission to Lucé, France, to study the circumstances of the fall of a stone, concluded, despite unanimous testimony of numerous eyewitnesses, *that the stone did not fall,* but that it was merely a terrestrial stone, which had been struck by lightning. An even worse case was soon to come. On July 24, 1790, another fall of stones occurred in south-west France. Many stones fell, burying themselves in the soil, and the very striking accompanying phenomena were seen by hundreds of people. No less than 300 written statements, many apparently sworn to, were sent in and pieces of the stone produced. The scientific journals did indeed print the accounts given, but only to ridicule the popular ignorance and credulity. The words of Bertholon, said to be quite typical, are worth quoting in the original:[7] "Que pourrions-nous ajouter à ce procès-verbal? toutes les réflexions qu'il suggère se présentent d'elles-mêmes au lecteur philosophe, en lisant cette attestation authentique d'un fait évidemment faux, d'un phénomène physiquement impossible."

But despite the verdict of the "philosophers," stones continued to fall, some in neighboring countries. Finally on April 26, 1803, near the village of L'Aigle in France, another shower of stones descended. By this time the confidence of the Académie had been shaken, and Biot was delegated to make an inquiry on the spot. The celebrated memoir containing his investigations finally proved to his colleagues that the stones did fall from the sky and were of cosmical origin. From that time to the present, with the exception of some fantastic theories advanced about the middle of the last century, which connected the smaller meteors with the weather or electrical phenomena, all meteoric bodies large and small generally have been

[7] *Pubblicazioni della Specola Astronomica Vaticana,* (II), **2**, 4, 3, 1913.

held to be of cosmical origin. The theories just mentioned apparently were never widely accepted and deserve no further comment.

The study of shooting-stars or meteors proper was first seriously undertaken in the year 1789. Nevertheless long before this time many eminent astronomers personally had believed in their cosmical origin. For instance, as early as 1686, when the appearance of a great fireball with a long terrestrial path and retrograde motion had called his attention to the subject, Halley made the supposition that in space there existed desseminated matter which is concentrated in its continual fall toward the sun, and produces, when it meets the earth, the phenomenon of shooting stars. But above all Chladni deserves the highest praise, among the older astronomers, for his clear vision and remarkable work in meteoric astronomy. In two memoirs which appeared in 1794[8] and 1819[9] respectively, he established on the basis of all the facts then known, the connections between meteorites and fireballs on the one hand, and between fireballs and meteors on the other. He further affirmed their possible connection with comets.

Leaving for the time the other opinions of Chladni, we find two young German students, who, carrying out his idea of corresponding observations of meteors made simultaneously in two places, really began practical meteoric astronomy. They were Brandes and Benzenberg, students at Gottingen. Between September 11 and November 4, 1798, they observed 402 meteors, of which they judged that 22 were identical, due to the meteors appearing at the same time as seen from both places and their similarity of magnitude and other characteristics. H. A. Newton gave the numerical results found for 21 of these meteors.[10] In all but 4 of the cases only the altitude of disappearance is given and these vary from 152 to 7 miles (both of these extreme values probably very erroneous), while the average comes about 61 miles. These figures gave the first approximate ideas as to the height of the stratum of the earth's atmosphere in which meteors appear and disappear.

Brandes carried out similar observations on an extensive scale in 1823, but this pioneer work seems to have attracted very little attention from astronomers, and meteors might have remained neglected

[8] *Ueber der Ursprung der von Pallas gefundenen und andern Meteormassen.*
[9] *Ueber Feuermeteore und ueber die mit denselben herabgefallen Massen*
[10] *Am Jour. Sci*, (II), **38**, 135, 1864.

indefinitely had not the great shower of November 12, 1833, attracted such immense popular attention that the subject could no longer be overlooked. From that date to the present the problems of meteoric astronomy, both practical and theoretical, have received the attention and study of many eminent men, and while from its very nature it is less susceptible of exact observation and prediction than most other branches of astronomy, yet very great progress has been made and work of permanent value has been done.

Before proceeding further it will be convenient to give a few necessary definitions The word meteor itself and its derivatives come from the Greek word *μετέωρον*, generally found in the plural *μετέωρα* which latter is used to describe any atmospheric phenomenon. We see this in our word meteorology, which is the science of the phenomena of the earth's atmosphere, and not the science dealing with what we now call meteors. As used at present the term meteoric astronomy includes what is known about all classes of bodies which, entering the earth's atmosphere from without, appear for a brief interval as a star-like object shooting across the sky, or, in rarer cases, larger bodies which shine with brilliant light, a few of which eventually reach the earth. Most writers divide such bodies into three classes: shooting stars or meteors, fireballs or bolides, and meteorites. These names will be retained in this book. Without attempting in this place to set up any theories as to their real differences in constitution or origin, it will here be sufficient to define their apparent differences on which the preceding classification rests. In general we will consider a meteor as a "shooting-star" which may vary in brightness from the faintest we can see to a body perhaps as bright as Saturn or Jupiter Fireballs generally are considered to be at least as bright as Jupiter or Venus, and in rare cases have been described as many times larger and brighter than the full moon, with all possible intermediate grades of brightness. The passage of such bodies through our atmosphere frequently is accompanied by sounds not unlike thunder. Often at the end of their path they explode. The term bolide usually is restricted to a fireball which bursts. No portion of a fireball is known to reach the earth except eventually as dust. Meteorites may give all the phenomena described as belonging to fireballs, or bolides, the only apparent difference being that meteorites reach the earth as solid bodies of various sizes, several hundreds of which are in our museums. While the foregoing classification admittedly is rather rough, it will serve our immediate purposes. Finally a "meteoric shower" is caused by the appearance on any given night of very large numbers of meteors, which seem to shoot across the sky in every direction outward from some given point or small area of the heavens, peculiar to that particular shower, and known as their radiant.

CHAPTER II

How Meteors Are Observed

Before going further into the history and theory of the subject the methods used for the observation of meteors will be described. Generally the observer depends upon the unaided eye. Furthermore, with rare exceptions, even the approximate region of the sky in which meteors may, at any moment be expected to appear, is wholly unknown. Besides, the meteor's visible span of existence usually is only a fraction of a second—perhaps one-half second on an average. Hence very accurate observations in the astronomical sense are quite impossible

What is desired most of all is the meteor's apparent path across the sky. Fortunately the stars serve as ready reference points and it is the observer's task to fix the meteor's course among them with all the precision at his command. The hour, minute and sometimes the second of appearance must also be known. It is obvious therefore that an observer must have at hand a watch, and either a celestial globe or a set of maps of the sky on which to plot the path. A good knowledge of the constellations is prerequisite for anything like quick and accurate work. As the whole phenomenon is quickly over in each case, it requires constant vigilance and considerable practice before one becomes proficient as an observer.

Any mechanical aid would be valuable for helping to fix the meteor's path, provided it could be used quickly enough Many observers find a simple, straight ruler or stick, from one to three feet long very helpful. The moment a meteor is seen the ruler is held at arm's length with its edge parallel to the exact path among the stars. This operation is greatly facilitated, of course, by any visible train. The direction having been noted in this manner, one attempts next to fix the points among the stars at which the meteor was first seen and finally disappeared. Chance is always a large factor in the accuracy with which this can be done. Generally the accuracy is highest in regions rich in stars. Each end point is fixed by mentally noting that it was on a line between two given stars, say half way between them or any other given fraction of their distance apart. Sometimes by noting that the path began or ended 1° or 2° in a given direction from some one star, or in some simple geometrical relation to a group of stars. A most valuable check is to note any star over which the path actually seemed to pass, or the fraction of the distance between two neighboring stars it passed from either on its way between them.

The above method must be used when corresponding observations are being made to determine heights. But if we desire only to determine radiant points, the usual aim when one is not working in conjunction with another elsewhere, certain simplifications can be made. In this case it is the precise direction which is needed rather than the exact points at which the meteor begins and ends. In other words an error of two or more degrees in these points as plotted, provided the direction is correct, will not alter the position of the derived radiant, unless indeed the path is very short, when such an error hardly could be made. Hence to plot the path any point on the path and the direction of motion are sufficient. It always is a far simpler thing to fix some one point than two given points, and it may be done by any of the methods explained. The direction can readily be determined by looking either forward or backward along the rod until we come to some star which will serve as a good reference point. If no star can be found, we must again estimate the fraction of the distance of our point between two stars on either side of the path from the one or the other. Turning now to our map, we have the two points necessary to draw the line representing the meteor's path. It goes without saying that one tries to make the indicated path of the same length in degrees as noted in the sky. An arrow head always is put on one end to show in which direction the meteor was going, and beside the drawn path a number, corresponding to that on the record sheet used for the night. The forms furnished to all members of the American Meteor Society are ruled as follows: At the head are spaces, with proper designations, for the name of the observer, place, hour and minute of the beginning and ending of observations, and the condition of the sky during the interval. The record proper is ruled in parallel columns so designated that they will contain respectively, concerning each meteor whose record fills one line of the page, its time of appearance, serial number for the night, class, color, magnitude, length of path in degrees, length of time visible, duration of train, right ascensions and declinations of beginning and ending points, remarks, serial number for the year, and, finally, the accuracy with which the observations were made on a scale of good, fair and poor.

To the uninitiated it may seem quite incredible that, with practice, the path can be drawn on the map and the whole record made in less than 60 seconds, with the exception of filling in the coordinates of the beginning and ending points of the path and serial number for the year, all of which can be done at leisure next day. With good fortune the record easily can be made in 40 seconds by an experienced observer

The actual observations having been briefly described, we must next explain what the radiant point of a meteor shower is, and how it can be determined from such observations To make this clear two figures are necessary, the one showing the cause of the phenomenon known as radiation, the second a copy of an actual map of the meteors which appeared in a certain area of the sky on August 11, 1921, as they were observed by the writer. The Perseid radiant is shown in figure 2.

In figure 1, let SN be the surface of the earth, the observer being at O. Let a stream of meteors, moving in parallel paths of equal length in the earth's atmosphere be represented by AA′ EE′. Let HH′ be the background of the sky, which geometrically is infinitely distant from O. Now project AA′. EE′ upon HH′ as seen from O. We find AA′ becomes aa′, BB′ become bb′, CC′ as it falls directly toward O projects into a single point c′, DD′ into dd′ and EE′ into ee′. We see at once that the angular paths aa′ . ee′ vary, and may be of any given length, within limitations of the figure. Also for instance aa′ is moving to the north, ee′ to the south. Yet if we project all the paths backward they will intersect in c′, which is called the radiant. It is obvious that other meteors XX′, etc., parallel to AA′ but not in the plane HH′SN, on projection would still pass through c′, though they would lie in other planes, which would of course contain Oc′. Meteors come at all possible angles of inclination to the earth's surface, but the proof is in no way invalidated if we incline AA′, etc., to SN by any other angle than that chosen in the figure.

A similar phenomenon is produced by looking down a railroad track which is straight for several miles. If we stand between the rails the portions near us seem to diverge one to the right, the other to the left, but in the distance they seem to meet Therefore the radiant point of a meteor swarm is simply the direction in space from which they appear to come, the point at infinity in which a given set of parallel lines appear to meet.

Looking now at figure 2, we see how observations can give us such radiants. This region contains the radiant of the Perseids, the meteors having been observed on August 11, 1921, at the McCormick Observatory, University of Virginia, by the writer Only a small part of the total number seen in the region were actually plotted that night for reasons needless to enter into here Of the 18 tracks

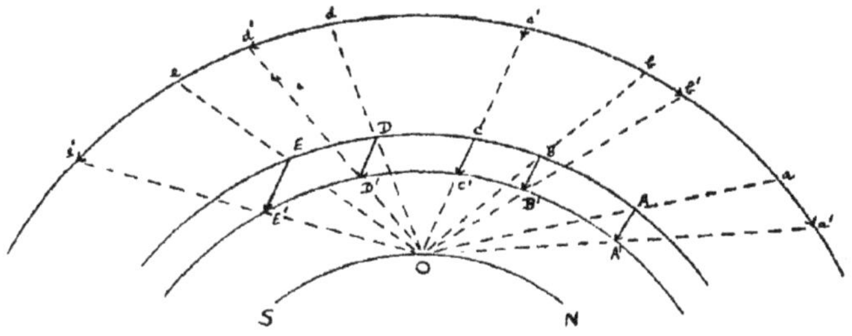

FIG 1

upon the map 16 are seen to be Perseids, but the very long track beginning almost at the radiant itself must have belonged to some other stream whose radiant was further to the south The numbers, which appear beside each arrow in the original record, have been omitted in the reproduction, having no meaning for this purpose. It should be understood that similar gnomonic maps, covering the whole visible hemisphere of the sky, were used on the night in question, but the one given here is quite sufficient for illustrative purposes. As most of the meteors plotted were quite near the radiant, it should have been well determined by such a number of tracks. That they do not all intersect absolutely in a point is due in part to errors of observation, but in this case also to actual non-parallelism of the tracks. Reasons for this latter fact will be given in their proper places. This method of determining the radiant

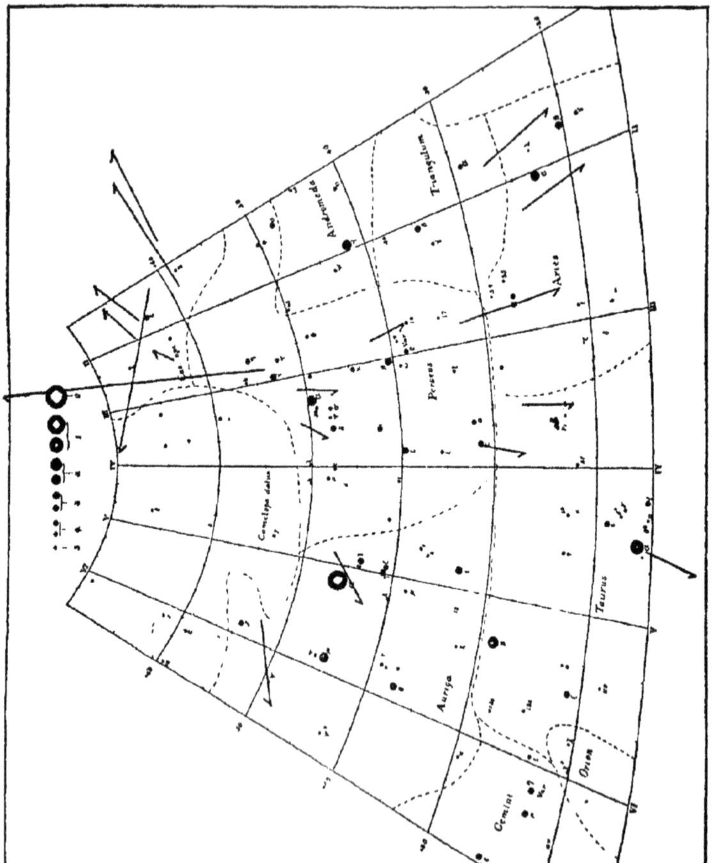
Andromeda
Triangulum
Aries
Perseus
Taurus
Auriga
Orion
Gemini

FIG. 2

directly from the plotted paths is known as the graphical. A more accurate determination may be made by the method of least squares, which naturally requires considerable computation. Several papers on this subject have been published. That by H. Chrétien in *Bulletin de la Société Astronomique de France,* 18, 484, 1904, is especially recommended because it is easily obtainable and also particularly clear.

This scarcely seems the place to go into fuller details of the practical methods for making observations of meteors. Persons interested may obtain circulars, maps, blanks and full instructions from the American Meteor Society, whose headquarters is at the Leander McCormick Observatory of the University of Virginia, or from circulars printed by other similar organizations. However, this much can be added. Experience has shown that from July to the end of the year many more meteors per hour can be seen than in the first six months. Further, that the later in the night the more frequently they occur. Taking the midnight hour to show about the average number and supposing that the observer has a moderately clear sky, during the first six months he might hope to see from 5 to 8 meteors per hour, during the second six months from 10 to 15. The causes of these variations will be discussed in Chapter XVI and more exact tabular data given on the rates per hour for each month.

In addition to naked eye observations, without further aid than noted, the brighter meteors may be photographed, if conditions are favorable. Also several simple forms of instruments have been devised which, used in naked eye work, have in some cases proved valuable in the hands of certain observers. Two such devices invented and described by the Rev. M. Davidson are especially worthy of mention.[1]

When it is realized that the best naked eye observations can give but approximate results for meteors, it is quite natural that photography should be pressed into service. Certain unsuccessful attempts to photograph meteors were made in Europe before 1890. Also, at various observatories a few impressed their tracks upon plates exposed to the sky for other purposes. Until very recently the only successful attempt to pursue this work regularly was made by Wm. L. Elkin, then director of Yale Observatory. His work began in 1893 and ended in 1909 During its continuance over

[1] *Jour., B. A. A.,* **30,** 92 *and* 223, 1919–1920.

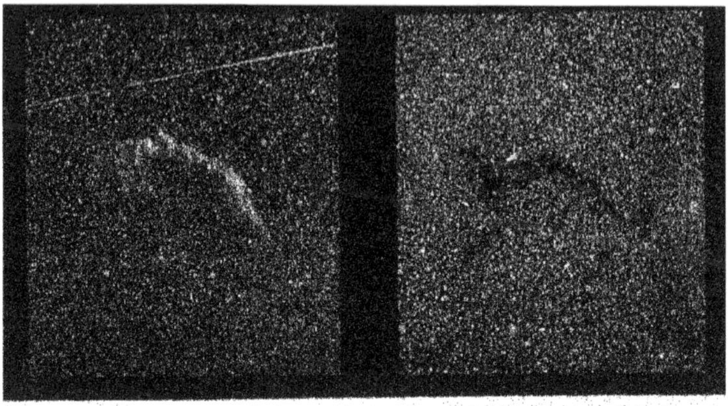

a, Gaseous nebula N G C 6995 and meteor trail Exposure 5^h 43^m, by E. E. Barnard

b, Dark object in Cepheus Taken July 15, 1909 by E E Barnard at Yerkes Observatory

one hundred meteors were photographed, almost entirely during August and November. Preliminary accounts of its progress, and of the instruments and methods used were published in three short papers which appeared in 1899 and 1900.[2]

In the first of these papers Elkin described the polar-axis, driven by clock-work, that was used to carry eight short focus cameras, with large fields of view. The camera lenses were 4, 6, and 8 inches in diameter, respectively, and had a focal ratio of about 1:4. Using 8 by 10-inch plates the cameras could cover an area of several hundred square degrees in good definition. The lenses were of the portrait type. At his second station a polar-axis, carrying four similar cameras, with lenses of about 5 inches diameter, was employed. Elkin later mounted a wheel with a number of opaque sectors before his cameras This was rotated at a known rate. Hence a meteor's photographed track had a certain number of breaks at regular time intervals. When the data for this track were combined with the data secured at the second station when a meteor had been photographed at both, it was possible to solve not only for the meteor's height and length of path, but also for its linear velocity. A few results of such calculations appeared in his third and last paper. While the writer is intimately acquainted with the details and final results of the whole work he is not at liberty to discuss them here further than to say that their immediate publication by Elkin would be of great service to astronomy, as results of value were obtained.

At Harvard College Observatory, from 1897 to 1904 inclusive, extensive observations of the Leonids were made. In all cases besides the visual observations numerous photographs were taken with many different instruments in hopes of securing data. Full accounts of the work are given in *H. C. O. Circulars* No. 31, 35, 40, 45 and 89, and *Annals* 41, Part 5. Some success was obtained in several of the years, and no less than thirty-four photographs of 11 different meteors were secured on November 14, 1898. However that night furnished the finest shower observed for the past thirty years with the exception of that of November 14, 1901. The general problem is also discussed in some of the circulars. It was found that only the bright meteors, i.e., those of at least second magnitude or brighter, have any chance of impressing their trails upon plates, as a general rule. Very recently an excellent radiant based upon

[2] *Astroph. Jour*, **9**, 20, 1899, **10**, 25, 1899, *and* **12**, 4, 1900

three Orionid meteors from plates taken on October 20, 1922, was given in *H. C. O. Bulletin* No. 778 and 783. The result of a search on plates, taken with the one inch Cooke lens of 13 inches focus is given in *H. C. O. Bulletin* No. 778. It is stated that stars of the eleventh magnitude are shown, that each plate covers more than 1200 square degrees, that the series extended for 23 years, and, that the average exposure time was 69 minutes per plate. In all 641 plates, with a total exposure time of 44,266 minutes, were examined. These plates were those taken during the periods August 1 to 15, October 16 to 21, November 10 to 20, November 15 to December 5, in other words including the dates of maxima of the Perseids, Orionids, Leonids and Bielids. The number of trails found for each epoch was respectively 3, 0, 1, and 8. The survey was equivalent to a photographic search for bright meteors for 738 hours over a region of 40° diameter. As is seen this resulted in finding only 12 meteor trails. Apparently this poor success was due to the focal ratio of 1:13. With such a ratio only very brilliant meteors could be expected to impress their trails upon the plates. This research, however, gives us the first scientifically stated results of what may be anticipated when a given lens is used with regularity over a long period of time

During the years 1920 to 1923 the writer experimented with several portrait lenses of 4, 5, and 6 inches aperture respectively, focal length about 1:4. These experiments were made in the months of May, June, August and October. The success obtained was moderate but quite enough to prove that such cameras may be expected to photograph any first magnitude meteor which crosses their fields, and with good fortune may catch meteors a magnitude or more fainter. If a meteor moves with low angular velocity or leaves a train visible for some seconds, its chance of impressing itself upon the plate is vastly increased. No data are yet at hand to show whether color plays any important rôle. The fastest obtainable 8 by 10 inch plates were used. The largest number of trails were secured on the nights of August 11 and 12. Using one 4-, one 5- and one 6-inch camera on the two nights, with about 9 hours total exposure for each, five trails were photographed, all by chance with the 6-inch. As the 1923 return of the Perseids was far from a rich, indeed rather a poor return, this proves that for these dates at least there is an excellent chance for photographing meteors with instruments similar to those mentioned. There is no

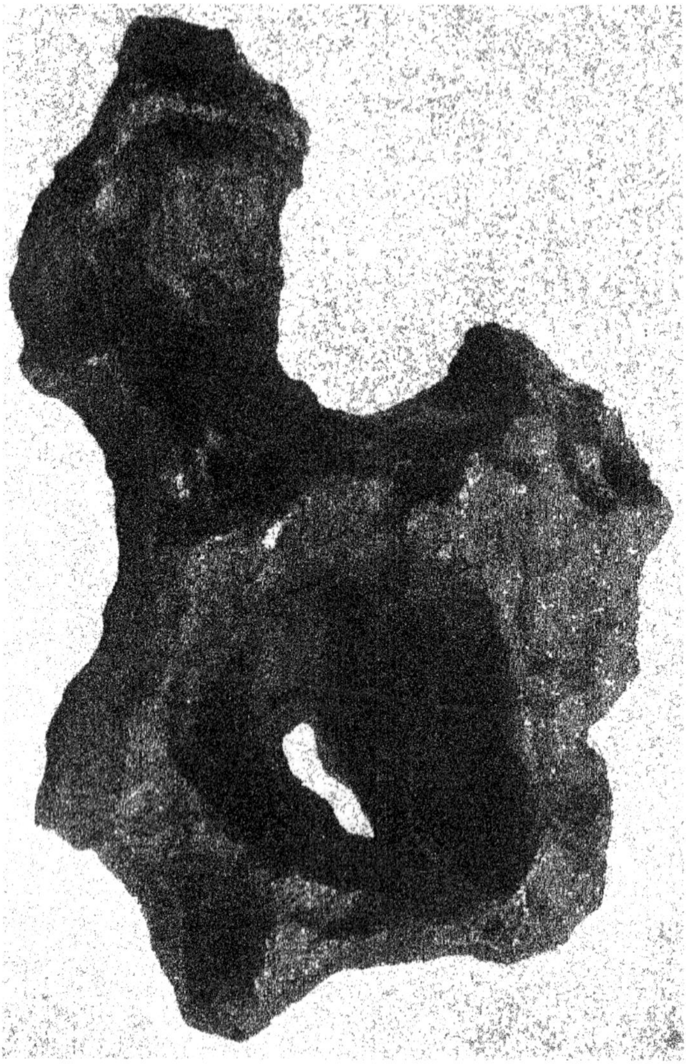

IRON METEORITE FROM COON BUTTE, ARIZONA

Now in National Museum, Washington, D. C. Photograph loaned by Geo. P. Merrill.

doubt that a regular campaign for photographing meteors, with the proper equipment of modern cameras, would yield a rich harvest of important data and very probably discoveries.

One of the latest known regular attempts to attack the problem has been made by F. A. Lindemann and G. M. B. Dobson in England. A full description of their apparatus, which has Goerz 25 cm. lenses focal ratio 1:3.5, has recently appeared.[3] A few words indicating that some success has been obtained in its use were added but the number of trails photographed was not given. The complete results of this investigation will be awaited with impatience.

A very interesting paper by J. Sykora has appeared just in time to be mentioned here.[4] In it he tells of many attempts to photograph meteors from 1901 to the present, during which 30 trails were obtained on his plates. He gives two splendid radiants thus determined for the Perseids, and most interesting data as to the heights at which a meteor of this stream burst twice, as well as where it appeared and disappeared. According to a visual observation of this same meteor it seemed to go much further than was shown upon the negative. If this turns out to be the usual thing, a most important fact has been discovered, one which will have serious bearing upon the development of the photographic method. Another most important point was that this meteor, on any reasonable assumption as to the duration of its visibility, had an enormously hyperbolic velocity.

It should be stated that the reduction of the position of a meteor's trail upon a plate is a matter which requires far more time and trouble than the reduction of the average naked eye observation. This is largely due to the fact that such a trail admits of precise measurement, and it is most desirable that the results be as accurate as possible. Practical limits of accuracy are indeed soon reached, but despite this all known methods of reduction are rather long. One complete published method by H. H. Turner occurs in *Monthly Notices, R. A. S.*, 67, 562, 1907. Numerous other articles refer to the photography of meteors, but apparently no other one dealing directly with the measurement of their trails upon plates has appeared in print.

[3] *Monthly Not*, *R. A S.*, **83**, 163, 1923.

[4] *Bul. Soc. Astr. de France*, **38**. 64, 1924.

CHAPTER III

COMETS

The intelligent discussion of certain phases of meteoric astronomy requires a clear conception of modern ideas of comets, particularly where they originated and in what manner their orbits vary under the planetary perturbations To the professional astronomer the material in this chapter is well known, but the latest researches of Stromgren and Russell have not found their way into the average text-book, hence for the amateur the following information will be valuable.

A typical comet is made up of three parts, nucleus, coma and tail. Some comets have only a coma, the other two parts apparently being lacking. In this case the center of the coma is the point which describes what we call the comet's orbit; if a nucleus is visible within the coma that is the point thus used. The coma contracts, contrary to every expectation, as it nears the sun, but the tail becomes more developed, the nearer the approach. The main tail always is pointed away from the sun, not exactly so but at a small angle with the radius vector of the orbit. The dimensions of the coma are immense, sometimes about as large as the sun itself when a great distance from it; others however are scarcely larger than the earth. We must infer that there are still smaller ones, which, owing to low density and poor reflective power, offer no chance of being discovered. In fact we have no present data by which to set a lower limit to a comet's size. The nucleus sometimes appears a few thousand miles in diameter, but it too varies remarkably, generally being smaller (apparently) when near the earth than further away. Doubtless irradiation plays some part here. The tail when most developed for a great comet may be many millions of miles long and at its wider end, which is away from the comet, may be several million miles across.

Theories of the constitutions of comets vary, particularly as to the mass involved. Many eminent astronomers claim the mass is almost nothing, that a comet could be "packed in a hat box" to borrow one's expression. The mass is known to be too small to be calculated by the method universally used, i.e., to see how much it deviates bodies near which it passes. Yet, at least so far as the writer knows, there is no particle of evidence for any such low mass as that mentioned above. Even if a comet weighed 10^{16} tons (the earth weighs 6×10^{21} tons) it still could not, so far as telescopic observations would show, perturb one of Jupiter's satellites appreciably. Of course, though, this observational failure would mean merely that the change in position was too small to be detected, not that the comet was a ghost without mass Another type of observation, however, has proved that the density of a comet's head is very low. On several occasions these objects have passed in front of some fixed star and the passage been carefully observed by trained astronomers with good telescopes. The usual verdict has been that the star showed little or no diminution in brightness, nor was it displaced by refraction effects.

The more usual idea is that the nucleus is made up of solid, discrete particles, doubtless at some distance apart, and all involved in a gaseous atmosphere which forms the coma. From these two, which together form the head of the comet, flows out the tail. We have no escape from the belief that nothing but gaseous matter constitutes the latter unless very near the head, because the volume is so immense that if enough solid particles were present to make the tail visible by reflected light, the mass could be detected.

A comet's spectrum proves that it shines both by inherent light and reflected solar light. The true cause of its inherent light is conjectural, and it is even more conjectural how the tail, almost infinitely more tenuous than our best terrestial vacuum, can shine so brightly. But it is no part of our purpose here to discuss the possible theories of this luminescence more than to mention that, among others, electrical, phosphorescent and radio-active effects have been suggested. It has however long since been accepted that the material which forms a comet's tail is expelled from the head by repulsive forces which act both from the head and also from the sun. Briefly, the material is ejected in any or all directions to a certain distance by the head itself, then reaching that limit at which the sun's repulsive force becomes the stronger it is all turned back and forced away from the sun, thus forming a tail. This matter is constantly renewed and is lost forever to the comet. Hence we may expect such an appendage to become fainter on successive returns. For comets of very short period it may scarcely exist, which doubtless means they have made many returns to the sun in orbits with relatively small perihelion distances. Masses of such material, in certain comets' tails, of peculiar shapes, can sometimes be traced for many days as they continually move further from the nucleus, with a velocity greater and greater the further they move away. This proves that the repulsive forces, at least in part, must be continuous in action.

According to Brédikhine there are three classes of tails, the hydrogen, hydrocarbon and iron. He supposed the ratio of the repulsive forces to gravitation to be from 12 to 15 to 1, from 2.2 to 1.1 to 1, and from 0.5 to 0.1 to 1 respectively for the three types. Some comets have at least two of these types, besides anomalous tails or jets on which latter his meteor theory is built (see page 207).

From what has been said we must look for solid material originally in the nucleus. Then by dispersion it may be found in the coma, and later beyond the latter's bounds. But under no possible theory can the usual comet generate meteors through the material it ejects to form its main tail. According both to theory and observation a comet is not a permanent body because it continually loses its material in several ways, without, so far as is known, offsetting any appreciable fraction of this loss by a corresponding gain. Of course it picks up a few meteors and possibly a little of the material of the tails of other comets, but we could not expect its gain to amount to as much as 1 per cent of the loss. It is not surprising, therefore, that many comets, even of short periods, disappear, leaving no trace behind them All the solar system must be populated by debris of comets, long since broken up into tiny fragments of their original great volumes

Do comets, then, come from interstellar space or are they original members of our solar system? The answer which seems complete has been given recently by Stromgren. It long had been a surprise to everyone that comets, assumed to come from space, do not show strongly hyperbolic orbits. In fact a half dozen or so orbits were catalogued as hyperbolic, according to some one computer or another, but in all cases their eccentricities barely exceed unity. In other words a parabola nearly satisfied the observations. Stromgren, in a series of masterly computations in which perturbations were allowed for, when the comets were on the outer confines of the solar system, proved that in every case the hyperbolic motion was due to some

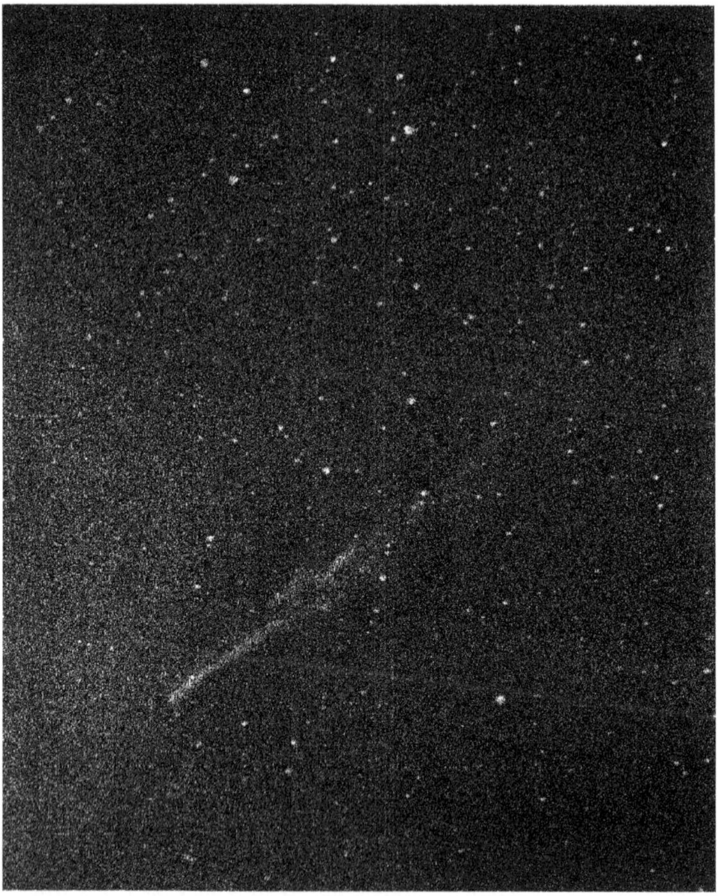

Brook's Comet, October 22, 1893
Photograph by Barnard at Lick Observatory

one or more of the planets. This cut down the classes to elliptical and parabolic orbits only. But it had long been admitted that the apparent parabolas could not be anything but ellipses with immensely long axes, hence his proof is considered complete that comets are veritable members of our system. True they are supposed to come from a great distance, perhaps 10^3 to 10^5 astronomical units, but nevertheless to accompany our system through space and to be a veritable part of it. According to the Nebular Hypothesis, it is supposed that residuals of nebulous matter, near the confines of the original nebula, at intervals move down toward the center. The same region could form several such comets in relatively quick succession, thus explaining the so-called comet groups. This home of the comets is frequently alluded to as a sort of nebulous shell at some such distance as that mentioned from the sun. It, however, remains to be explained how the nucleus of a comet could develop under such conditions.

In every text-book on astronomy up to the last three years the statement will be found that each of the major planets has what is denominated a comet family. In 1920, H. N. Russell of Princeton studied the question,[1] basing his work upon an earlier conclusion of H. A. Newton that the number of comets of which the period is inferior to P is proportional to the $\sqrt{P}$ Without entering into his arguments, he concluded that if the theory of capture be admitted Jupiter must have done nearly all the capturing. Saturn could account for only one in forty, Uranus and Neptune together for only 1 in 4000. He then notes that the families attributed to all but Jupiter are far too numerous to fit the theory. Also that a comet may have the same aphelion distance as a planet, but due to the different inclinations of the orbits the two aphelia may be very far apart. Thus for the seven members of the so-called Neptune family, no one comes nearer than 4 astronomical units to the orbit of the planet. No sound reasons remain to believe in the families of Saturn and Uranus. If indeed Saturn is partly responsible in any given case, the later effects of Jupiter would have masked the earlier of Saturn. He concluded that the elements of the periodic comets satisfy the laws calculated on the hypothesis of capture and thus receive strong support, but that his study overthrows the former

[1] *Astron Jour.*, **23**, 49, 1920.

ideas as to the origin of the perturbation which led to the capture. A partial objection to the conclusion drawn by Russell that the families of the three outer planets do not exist as such has been advanced. Briefly stated it is that the comets may have been captured by the planets in question and their orbits so changed by later perturbations of other planets that at present their planes are so inclined that they are quite distant from the orbit of the capturing planet.

Opinions later expressed on the connection of comets and meteors, the formation of meteor streams, etc., are based upon the results given in this short discussion of the best available theories.

CHAPTER IV

The Leonids

The early progress of meteoric astronomy is so intimately connected with the appearances of the great Leonid showers that it can best be traced through a study of these phenomena themselves and of the stream that caused them. As has been mentioned it was the shower of 1833 that was the cause of the sudden interest in the subject. On November 11, 1799, a similar shower appeared and was fairly well described by a few competent witnesses It is best known from the description given by the great traveller Humboldt, who happened to be in Cumana, South America, at the time. It seems to have been visible over at least 90° in longitude and 64° in latitude of the earth's surface, its greatest brilliancy occurring in the western hemisphere. In Germany in some places, however, many meteors were seen, and doubtless had proper records been kept, at least a fairly bright shower would have been recorded for all western Europe. Humboldt's account stated, in part, that thousands of meteors and fireballs were visible moving regularly from north to south, all meteors leaving streaks 8° to 10° long, and lasting 7^s to 8^s. That there was no part of the sky so large as twice the moon's diameter that was not filled each instant by some meteor or fireball. That the fireballs often burst, the largest 0.5° to 0.6° in diameter consumed away without leaving sparks, but leaving bright trains behind them lasting from 15 to 20 minutes. The light of the meteors was white. He saw the phenomenon for the space of four hours, daylight only putting an end to it. A few were even seen a quarter of an hour after sunrise. However the maximum had been passed about 16^h (i e., 4·00 a.m., November 12), not so many appearing after that hour as before. Inquiry among the natives elicited the information that in 1766 a similar shower had been seen.

One other account,[1] by Andrew Elliot, then en route from Philadelphia to New Orleans, may be quoted in part:

November 12, 1799, about three o'clock a.m. I was called up to see the shooting of the stars (as it was commonly called). The phenomenon was grand

[1] *Trans Am. Philos Soc.*, **6**, 26, 1804.

and awful, the whole heavens appeared as if illuminated by skyrockets, which disappeared only by the light of the sun after daybreak The meteors which appeared at any one instant as numerous as the stars, flew in all possible directions, except from the earth toward which they all inclined more or less, and some of them descended perpendicularly over the vessel we were in, so that I was in constant expectation of their falling among us. We were in latitude 25° N. and S.E from Kay Largo, near the edge of the Gulph Stream. . . . . I have since been informed that the phenomenon extended over a large portion of the West India Islands and as far north as Mary's in latitude 30° 42′ when it appeared as brillant as with us off Cape Florida

The account of the wonderful phenomenon of 1799 passed more or less unnoticed until 1833, when the great shower returned on the night of November 12. It should, however, be remarked that meantime on November 13, 1831, and again on November 12 and 13, 1832, unusual numbers of meteors were noted by ship captains at sea or observers on the Continents of Europe and Asia It is certain indeed that a splendid shower was seen in Asia in 1832. This shower of 1833 also appeared in its greatest brilliancy in North America, where there were men of ability who not only made desultory notes but with all the means at their disposal carried out scientific observations. Meteoric astronomy really began with this shower.

This return of the Leonids was so remarkable that the impress upon the popular mind has never been obliterated and the interest then aroused in meteors has never died out. In *Silliman's Journal* **25,** 354–411, and **26,** 132–174, may be found most complete accounts of what was seen in America. It will be of interest to quote in part certain of these accounts by eye-witnesses. The first is that of Prof. Denison Olmsted, of Yale College.

About daybreak this morning our sky presented a remarkable exhibition of fireballs, commonly called Shooting Stars The attention of the writer was first called to this phenomenon about five o'clock, from which time until nearly sun rise, the appearance of these was striking and splendid, beyond anything of the kind he has ever witnessed

To form some idea of the phenomenon, the reader may imagine a constant succession of fireballs, resembling rockets, radiating in all directions from a point in the heavens. . . . . They commenced their progress at different distances from the radiating point but the lines they described, if produced upwards would all have met in the same part of the heavens . . . the balls . . . . just before they disappeared exploded . . . . No report or noise of any kind was observed . . . . [there were] meteors of various sizes and degrees of splendor. some mere points but others were larger and brighter than Jupiter or Venus, and one was nearly as large as the moon.

The flashes of light were so bright as to awaken people in their beds. . . . . A quarter before six . . . . it occurred to the writer to fix its place [i.e., the radiant]. . . . . During the hour following the radiating point remained stationary in the same part of Leo, although the constellation in the mean time . . . . moved . . . nearly 15°. . . . .

An observer from Boston, Mass., gave the following rates derived from his observations. In about one-tenth or less of sky from 5:45 to 6:00 a.m., 650 were counted. Making allowance for those lost in counting, 8660 would have been visible during the interval. At six the phenomenon was beginning to cease Unfortunately this seems the only rate actually counted by any of the observers, many of whom made excellent observations in all other respects. However, H. C. Twining at West Point, New York, said: "I should not deem it extravagant to suppose ten thousand to a single hour, during the period of my observations " These latter began just after 5:00 a.m. Further south the maximum is definitely put between 2:30 and 4:00 a.m.

The following quoted in part from the *Georgia Courier* describes the shower as seen by an observer near Augusta, Georgia, far south of New England:

. . At about nine p m the shooting stars first arrested our attention, increasing in both number and brilliancy until 30 minutes past 2 a.m., when one of the most splendid sights perhaps that mortal eyes have ever beheld, was opened to our astonished gaze. From the last mentioned hour until daylight the appearance of the heavens was awfully sublime It would seem as if worlds upon worlds from the infinity of space were rushing like a whirlwind to our globe . . . and the stars descended like a snow fall to the earth . . Occasionally one would dart forward leaving a brilliant train which . . . . would remain visible, some of them for nearly fifteen minutes . . .

More or less similar accounts are given by observers in New York, Pennsylvania, Ohio, Maryland, Virginia, North Carolina, South Carolina, Mississippi, Louisiana, and Missouri, and from ships at sea off the Atlantic coast. The phenomenon caused immense terror among the ignorant and superstitious, many believing that the "Day of Judgment" had come. The writer has often had the old negro cook of his family, who was born and raised in Albemarle County, Va., give him a vivid account of how she saw "the stars fall," when a girl of ten—incidentally the only means of fixing the approximate date of her birth. Though the shower occurred sixty or seventy years before, the impression never left her which had been made on her mind when a child, and she vividly described the terror of the negroes as "the stars fell, and fell, thick as snow coming down in a snow storm," and of how all thought "the Day of Jedgment had sho' come."

The fact that the meteors radiated from a fixed point or small area was plainly stated by, in addition to Olmsted, A. C. Twining of West Point, N. Y., W. E. Aikin at Emmittsburg, Md., V. H. Barber of Frederick, Md., J. L. Riddell of Worthington, Ohio, and, less clearly, by J. N. Palmer near New Haven, Conn. Also Captain Parker of the ship *Junior* and Captain Seymour of the ship *De Witt Clinton* observed the radiation phenomenon but the positions assigned by them to the radiant were considerably in error, while those of the other observers were quite nearly correct and coincident. The fact that this point was near the zenith, when the shower was observed by persons who saw it just before daybreak, caused others to believe that it *was* the zenith. All these names are cited because in nearly every book the whole credit is given to Olmsted and Twining, absolutely no mention being made of the others who made exactly the same discovery. This doubtless is owing to the fact that these two following up their discovery by further excellent work on meteors, which indeed laid the foundation of the science. To Olmsted is due the further credit of collecting and publishing the work of all the others mentioned, and we may justly hold him as being the only American, except his later successor, H. A. Newton, who during the nineteenth century deserves permanent recognition for his eminent contributions to meteoric astronomy. His conclusions on the 1833 shower will be briefly reviewed, but it must be remembered that he was a pioneer making the first investigations on a wholly new subject of study.

They were: (1) That the meteors of November 13 had their origin beyond the limits of our atmosphere. (2) The height from which the meteors emanated, above the surface of the earth, was about 2238 miles. (3) The meteors fell toward the earth attracted by the force of gravity. (4) The meteors fell in straight lines and directions, which within considerable distances, were nearly parallel with each other. (5) The meteors entered the earth's atmosphere at a velocity equal to about four miles per second. (6) The meteors consisted of combustible matter and took fire and were consumed in traversing the atmosphere. (7) Some of the meteors must have been bodies of great size. (8) The meteors were combustible bodies and were constituted of light and transparent materials. Further he concluded that the nebulous body which furnished the meteors: (9) was pursuing its way along with the earth around the sun; (10) that the body revolves around the sun in an elliptical orbit, but little inclined to the plane of the ecliptic, and having its aphelion near the orbit of the earth; (11) that the body had a period of nearly six months, etc.

Briefly conclusions (1), (4) and (10) which proved absolutely correct, with (3) and (6) which may be considered to be partly correct, are the basis of all subsequent work on the subject. His erroneous conclusion (2), along with (5) and (11) which in part follow from it, was due to his taking as actual differences the mere accidental errors in the determination of the position of the radiant made by the various observers mentioned and then using these observations to determine a rough parallax. This procedure seemed the more justified to him as he learned that the shower was not seen as a brilliant one in Europe nor indeed south of the equator. He therefore reasonably, if erroneously, concluded that had it come from a very distant radiant it necessarily would have been seen from the whole earth. Even today there is room for argument as to the size and constitution of meteors, particularly fireballs, hence no comment will here be made on (7) and (8). That the period was half a year, i.e., (11), was based further on the fact that he had learned that in Arabia a very fine shower had been observed in 1832 on the same day of the month. His reasoning was sound had his assumptions been so, i.e., had the same group of meteors furnished the shower in the two succeeding Novembers. What was really the case will appear in due time Olmsted's memoir further proves that neither the wind nor magnetism were causes, and he went so far as to call the "nebulous body" a comet, all of which proves that his grasp of the subject was most remarkably broad and correct. He states that his conclusions are concurred in by, and were made by the aid of, A. C. Twyning who thus justly shares the credit. In a later paper he refers to the "nebulous body" as a "meteoric cloud" and having meantime learned of a partial, if very inferior, recurrence of the shower in 1834, his opinion upon its semi-annual character was fortified.

Unfortunately it seems to have been considered necessary for the acceptance of any such hypothesis as that of a "meteoric cloud" that it should not only account for the November meteors but also for showers seen at other times of the year. This preconception led to much confusion and unnecessary work. The erroneous belief in a very short period for the Leonids was largely held until 1866, the fact that the Perseid meteors also gave an annual display, remarked by 1839, tending to confirm the belief. It is true that on most years only a few Leonids were seen, but always some, and the ready explanation was that the "meteoric cloud" was perturbed by Mercury and Venus so that in these years the earth only traversed the outer parts where meteors were sparsely scattered. However, Olbers believed rather that, instead of one swarm, there were several moving in the same orbit, but with long intervals between them, which respectively furnished the 1832, 1833 and 1834 meteors. He thought 3 or 6 years, or even 34 years might be the real period; attention having meantime been focused upon the 1799 shower in connection with that of 1833. He writes in Schumacher's *Jahrbuch* for 1837 as follows: "Perhaps we shall have to wait until 1867 before seeing this magnificent spectacle return." Herrick too, in 1839, affirmed that the period was 33 or 34 years, having looked up in old chronicles many accounts of meteoric showers back to 686 B.C., some of which fit in with such an hypothesis. Olmsted and Twining considered also the chance of a return in 1867.

The fact that the shower came in 1799 on November 11 and in 1833 on November 12 could be explained by a precession of its node. Finally Boguslawsky found a value for this by assuming that the shower of October 21, 1366, O. S. was the same as that of 1799 and 1833 He thus found an annual displacement for the point of meeting of the earth and meteor shower of $+1.835'$. Later learning of the shower of 845 October 16, O. S., and combining with those of November 11, 1799, November 12, 1833, and November 13, 1839, he found a retardation of 22 days in 994 years, or an annual node variation of $+1.5'$. Humboldt in his *Cosmos* from this concluded that the swarm must move in a retrograde direction in its orbit. During the next 20 or 30 years numerous astronomers took up the problem of trying to find coincidences of the several meteor showers, by then recognized, with accounts in ancient documents.

Mention will be made of certain of these cases later.

H. A Newton in 1863 published an important paper[2] on the subject which was inspired by a desire to prove the cosmical origin of meteors, a fact brought again into dispute by Quételet in his *Physique du Globe.* He sets up this proposition if meteor showers are due to anything within the earth's atmosphere they ought to return on the same date, i.e., follow the tropical year; but if they are cosmical in origin they should follow the sidereal year. As the latter is 0 014 day longer than the former, if meteors are cosmical then they ought to return one day later every seventy years. He reduced the longitude of the earth, at the date of showers, to what it would be in 1850, and then expressed the date as if in 1850. Neglected terms could amount to only ±0 7 day. For the Leonids he found:

A D.	585 Oct 25	corresponding to A D. 1850 Nov.	12 3
	902 Oct 29 or 30		11.0-12.0
	1582 Nov 7		10 7
	1698 Nov. 8 6		11 6
	1799 Nov. 11.6		12 9
	1833 Nov. 12 7		13.3

He followed this with a further table, containing many showers in October and November, and made the remark that by giving the November ring a precession of one day in 70 years most of these would be brought into the November period. In a later paper[3] he gave the actual copied accounts of those showers which he considered to have been the Leonids. They were in A.D. 902 (some confusing as to whether it could have been 901 or 903), 931, 934, 1002, 1101, 1202, 1366, 1533, 1602, 1698, 1799, 1832 and 1833. The accounts were gathered from Chinese, Arabic and European annals, and the article is one of great interest, as well as importance. In continuation[4] of the former paper he gave the table on page 30 which is copied in full.

[2] *Silliman's Jour.* (II), **36**, 146, 1863
[3] *Silliman's Jour*, (II), **36**, 377, 1863.
[4] *Silliman's Jour*, (II), **37**, 53, 1864.

NUMBER	A D	DAY		HOUR	LONGITUDE	a − nt	DIFFERENCE	END OF CYCLE	DIFFERENCE	PERTURBATION
1	902	October	12	17	24° 16 6′	24° 18 1′	−1 5′	901 50	+0 50	−238
2	931		14	10	25 57 5	25 7 7	+49 8	904 75	−3 75	+497
3	934		13	17	25 31 6	25 12 8	+18 8	934 75	−0 75	+467
4	1002		14	10	26 44 8	27 9 2	−24 4	1001 25	+0 75	+366
5	1101		16	17	30 2 4	29 58.6	+3 8	1101 00	0 00	+126
6	1202		18	14	32 25 5	32 51 4	−25 9	1200 75	+1 25	+622
7	1366		22	17	37 47 9	37 32 0	+15 9	1367 00	−1 00	−621
8	1533		24	4	41 11 7	42 17 8	−66 1	1533 25	−0 25	−048
9	1602		27	10	44 18 9	44 15 9	+3 0	1599 75	+2 25	−381
10	1698	November	8	17	47 20 6	47 0 1	+20 5	1699 50	−1 50	−269
11	1799		11	21	50 1 6	49 52 9	+8 7	1799 25	−0 25	−146
12	1832		12	16	50 49 0	50 49 4	−0 4	1832 50	−0 50	+037
13	1833		12	22	50 49 5	50 51 1	−1 6	1832 50	+0 50	+316

In this article it was found that the annual period is 365.271 days, and that the cycle is about one-third of a century. This was computed more exactly to be 33 25 years. In the table, after the columns containing the time of occurrence, was given, in the fourth, the longitude of the earth at that date. In the fifth column a = 51 17 7′, n = 1.711′, and t is the number of years to 1850. The sixth column is the difference between the fourth and fifth columns, which is merely the distance of the earth from the mean position of the node at the dates the showers appeared as given in second and third columns. As an error of one hour corresponds to 2.5′ in this column, quite a large part of the residuals are certainly due to the fact that the exact hour of maximum was never known The eighth column proves that fine displays may be expected from one to four years after or before the actual or computed time of maximum The last column gives the perturbations in the length of the earth's radius vector, the largest given being about 9000 miles. Newton hence concluded: (1) That a ring of uniform density around the sun did not properly represent the Leonid group, because the perturbations of the earth's radius vector were not sufficient to throw the earth in or out of the group far enough to allow for the results observed. (2) That a ring of very unequal density best represents the group, the densest part being concentrated within a fraction of the whole circumference.

(3) The motion is retrograde as the longitude of the node always increases. (4) The periodic time must be limited to five values; therefore in a year the group must perform $2 \pm \frac{1}{33.25}$, $1 \pm \frac{1}{33.25}$, or $\frac{1}{33.25}$ revolutions. Although Newton could not decide which was the correct period, and leaned toward $1 - \frac{1}{33.25}$, yet he predicted that November, 1866, would probably see a great shower, indicating further that a year before or after that date showers might also be expected. He, however, stated that the correct period could be decided upon once the true motion of the node was determined.

It will be further seen that if the period of $1 - \frac{1}{33.25}$ years were correct, the meteors' velocity could not be very different from that of the earth. Allowing for the earth's attraction he calculated that the relative velocity would be 20.17 miles per second, with the above period. We can see at once that had there been any accurate way at the time of determining the velocity of a Leonid meteor by observation, the theory of a short period would at once have been disproved, the relative velocity being, as we know, about 44 miles per second. S. C. Walker in 1841[5] had indeed calculated an orbit, but while he too found the motion retrograde, having also used too small a velocity, he reached the absurd result of a period of 0.356 year, and other elements hopelessly erroneous. The small velocity used by him was taken from a summary of all available observations. It seems that the earlier observers generally overestimated the duration of time a meteor was visible—a habit noticeable at present in every beginner in the subject. This consistently gave them low observed velocities. Walker's article is worth noting, however, because of an excellent summary of the work of others and because he devised an independent method of computing the orbits of meteor streams. In 1866, 1867 and 1868 fine showers of Leonids were seen over various parts of the earth, thus completely fulfilling the prediction of Newton (and the intelligent guesses of others already mentioned).

In 1866 appeared the work which from that date to the present has been the standard book on meteoric astronomy, *Note e Riflessioni*

[5] *Trans. Am. Philos. Soc.*, (N.S.), **8**, 87, 1843.

sulla theoria astronomica della Stelle cadenti by J. V. Schiaparelli, director of the Royal Observatory of Milan. This was translated, in an amended and enlarged state, into German in 1871 under the title *Entwurf einer Astronomischsen Theorie der Sternschnuppen* by George von Boguslawski. As the German edition is the one usually to be found in libraries and as it will henceforth be necessary to refer very often to this work it will simply be referred to as *Sternschnuppen,* with the page or article in the German edition.

In the original papers, published in Italian, which formed the basis of the book the connection of the Perseid meteors with Comet 1862 III was proved by Schiaparelli, and an orbit for the Leonid meteors was given. A little later, in 1867, within five days the identity of the Leonid orbit with that of the Comet 1866 I was discovered independently by C. F. W. Peters, Schiaparelli and von Oppolzer in the order named. But to Schiaparelli goes the credit for having earlier calculated the orbit of the stream, which made the further discovery possible. This orbit was based upon a period of 33 25 years, which he believed for many reasons to be the correct one.

Early in 1867, before the discovery of this identity, Le Verrier had published a paper[6] in which, assuming that 33 25 was the proper period for the Leonids and that the middle of the period fell at 1866.75, then going back 52 periods = 1729 years, he reached 137.75 A.D. He stated that the discontinuity of the phenomenon proved that they did not form a uniform ring, but rather a group at a certain point, yet a group of considerable length. He then argues that the perturbations of the various planets must tend to break up the main group and spread it along the whole ring. Therefore as time goes on the group becomes more scattered and hence less dense. Eventually it would be almost uniformly scattered around the ring as was the case with the Perseids. This means that each recurring shower should be less remarkable than the last until in the end no conspicuous shower could appear from the group. As this was not yet the case Le Verrier felt certain that the Leonids were recent comers into our system. With P = 33.25 years he calculated elliptical elements. Going back step by step, using his elements, he believed that he eliminated the other possibilities until he found Uranus, in 126 A.D., in such a position that it could produce the necessary

[6] *Comptes Rendus,* **64,** 94, 1867

perturbations upon a swarm moving in a parabolic orbit. To make this accord he had to arbitrarily change the node of the swarm in 126 A.D. by 1.8°, and the longitude of perihelion 4°. He felt that such changes were justified, as being within errors of observation. He further made the supposition that originally the swarm might have been spherical and perhaps moving in a direct orbit, either parabolic or a long ellipse. The action of Jupiter in 1770 upon the comet of that year was quoted as showing how vastly the orbit of such a body could be modified at a single passage by a major planet.

Shortly after this the identity of the Leonid orbit with that of Temple's Comet (1866 I) having been discovered by others, Schiaparelli was able to deduce more exact elements than those of Le Verrier or indeed those of his own original orbit. From this information and other considerations he argued[7] that the mass of Uranus was too small to have turned the stream out of an orbit differing much from the present one without at the same time having scattered the group. The perturbing body, which could cause the change from a parabolic to a long elliptical orbit, was then either Jupiter or Saturn.

The following are the two orbits on which Schiaparelli based his proof, as given in *Sternschnuppen*, p. 57:

	LEONIDS	COMET 1866 I
Perihelion passage	Nov 10.092	Jan 11.160
Longitude of perihelion	56° 25.9′	60° 28.0′
Ascending node	231 28.2	231 26.1
Inclination	162 15.5	162 41.9
Perihelion distance	0.9873	0.9765
Eccentricity	0.9046	0.9054
Semi-major axis	10.34	10.344
Period	33.25	33.176

Schiaparelli in his book, page 56, complains rather bitterly that his earlier work was nearly unknown in France and hence that the wide circulation of which Le Verrier's work was assured tended to throw his own somewhat into eclipse. Whatever just grounds he then had for complaint have, however, long since been removed and his work as a whole stands preeminent in meteoric astronomy.

[7] *Les Mondes*, **13**, 501, 1867; also *Am. Jour. Sci.*, (II), **44**, 129, 1867.

In April, 1867, J. C. Adams of England, basing his work[8] on Newton's suggestion that the motion of the node would give the true period and with the information furnished by the return of 1866, calculated the progressive increase due to the perturbing influences of Venus, Earth and Jupiter. Their joint effect for the four shorter possible periods derived by Newton were in no case more than 12′ in 33.25 years. However, the motion observed was 102.6″ annually with respect to the equinox or 52.6″ with respect to the stars, which equals about 29′ in 33.25 years. Adams now calculated the perturbations, assuming this latter period, produced by Jupiter, Saturn, and Uranus and found it collectively to be 28′. This excellent agreement—all that could possibly be expected in the case—forever settled the question in favor of the longest period of the five, i.e., 33.25 years. Adams elliptical elements of the swarm follow:

Radiant (1866) . .	R A 9^h 56^m Decl $+23°$ 1′
Period (assumed)	33.25 years
Mean distance . .	10.3402
Eccentricity	0.9047
Perihelion distance	0.9855
Inclination	163° 14′
Longitude of node	51 28
Longitude of perihelion	58 19

It will be noted that at perihelion the meteors are very near indeed to the earth's orbit, and the angular distance from perihelion to the point in which they meet the earth is only 6° 51′.

As all astronomers had fair warning, the return of November 13, 1866, was fully observed. In England the maximum occurred about one hour after midnight and was over by four o'clock. Dawes counted, with one assistant, 2800 meteors, from midnight to 14^h 13^m 10^s, in the eastern half of the sky. At Greenwich, the whole sky being covered by eight good observers, 8000 were counted, 4860 between 13^h–14^h.

In 1867 observations in England and at the Cape of Good Hope were quite negative, so far as any real shower was concerned. In North America brilliant showers were seen across the whole continent and from Canada to Mexico.[9] At Toronto, from 16^h to 17^h, 1345

[8] *Monthly Not. R. A. S.*, **27**, 247, 1866–7.
[9] *Am. Jour. Sci.*, (II) **45**, 78, 1868

meteors were counted; at Chicago, from $15^h\ 30^m$ to $16^h\ 12^m$, 1529 were counted; and other places gave similar rates. At New Haven, Conn., H. A. Newton estimated that the hourly rate after midnight was 900, and, as the moonlight caused great numbers to be lost, from 10,000 to 20,000 would have been visible but for that cause. In 1868 the shower was again well seen in America. On November 13, from 11^h to $18^h\ 11^m$, (40 minutes break, some clouds and an irregular number of observers participating) at Bloomington, Indiana, Prof. D. Kirkwood[10] reported 3280 meteors counted, the maximum being at 15.5^h when 900 were seen in preceeding 45 minutes Quite a number had been seen (165 in 3 hours) the night before Other similar records could be quoted.

From the brief accounts given it may be seen that for three successive years brilliant showers indeed came, but no one of them comparable to intensity to that of 1799 and 1833. Nevertheless they were wonderful, and vividly impressed both scientists and public, and Newton's prediction was brilliantly fulfilled. During all the period until 1899 it was confidently expected by everyone that the earth would again pass through the main part of the stream and that another great shower would occur, perhaps also on successive years as in the case of the previous return.

Astronomy having meantime progressed, demand for greater accuracy was made and two English astronomers, G. H. Stoney and A. M W. Downing,[11] undertook the laborious task of computing what had happened to the Leonids from 1866 on, due to the perturbations of various planets. J. C. Adams in his work just reviewed on page 34 had computed the perturbations of the node, and correctly determined which was the real period. They first showed that he used Gauss's method which among other things presupposes that the planet and perturbed body have incommensurate periods, so that in time they take every relative position with regard to one another. This condition for the Leonids is most imperfectly fulfilled, as fourteen revolutions of Jupiter equals almost exactly five revolutions of the Leonids, nine of Saturn equals eight, and two of Uranus equals only a little more than five. During the 1000 years over which we have satisfactory records, used by Newton and Adams as a basis for the latter's research, these cycles have been several times re-

[10] *Monthly Not R.A.S.*, **29**, 62, 1868.
[11] *Proc. Royal Soc.*, **64**, 403, 1898–9.

peated. This caused oscillition in the rate of advance of the node from its mean value, sometimes one way, sometimes the other. In 1533 A.D., for instance, the shower was 26 hours ahead, and similar changes are to be expected in the future. The perturbations have not only differed in different revolutions, but change greatly in a single one, due to the changing relative position of the planets.[12] The authors then introduced two terms. ortho-Leonids which travel in nearly identical orbits and form a compact stream which takes nearly three years to pass a given point of its orbit, and clino-Leonids which pursue orbits somewhat different and give us the small annual returns and those seen on days before and after a great maximum. The earth takes only five or six hours to pass through the ortho-stream, even though it goes through very obliquely. The clino-stream is very much wider and less dense, as is obvious. The authors being chiefly interested in the ortho-stream that alone was considered. They showed that both sinuosities and unequal densities occur at different parts of this long stream.

They further took that part of the ortho-stream for which Adams had computed the orbit, calling it segment A, and extended the perturbations to January, 1900, when exactly that same part of the stream would reach the earth's orbit which had given the showers of November, 1866, i.e., 33.25 years before. Adams' orbit was approximate, being based only on naked eye observations and not having certain corrections to the observed radiant properly included. However any small error in the assumed orbit would have no appreciable effect on the perturbations calculated. These latter were calculated by the method of mechanical quadratures for the effects exercised by the planets Mars, Jupiter, Saturn, and Uranus. Those of Venus and Earth were found to be insensible. The most noteworthy features were near approaches to Saturn in 1870, and to Jupiter in August, 1898, when the part of the stream in question, segment A, came within 0.9 astronomical units of the latter planet. Both of these planets therefore produced large perturbations. The orbits need not be given here, but the noteworthy change was in the perihelion distance which decreased from 0.9855 to 0.9729, thus throwing the stream further within the earth's orbit. Another result was that the period increased by one third of a year. As segment A

[12] *A. Berberich, Astr. Nach*, **147**, 359, 1898.

COMET 1908 c (MOREHOUSE) ON SEPTEMBER 13, 1908, AT $14^h 22^m$ C S T. Exposure $1^h 56^m$, by E E. Barnard at Yerkes Observatory

could not reach the earth's orbit before January, 1900, at earliest, they finally concluded, calling segment B that which would be passed through by the earth in November, 1899; (1) if segments A and B were in 1866 moving in orbits which differed little; (2) if segments A and B had in the interval since suffered practically the same perturbations, then a shower was to be expected November 15.75, 1899, G. M. T. As neither assumption was certain, though thought to be probable, the prediction was made by the authors with reservations. Continued investigations by Stoney during 1899 made him more skeptical of a certain return, and on November 10, 1899, before the Royal Astronomical Society, he gave further data showing that on the critical date Adams' orbit was 1,300,000 miles nearer to the sun than where the earth would be. Hence the only real hope for the shower was that the stream would be wide enough to reach the earth anyhow (or that due to imperfect data the calculations might somehow have been wrong).

Returning now to the actual events, for several years previous to 1899 the Leonids were carefully observed, but up to and including 1897 nothing but scattered members of the swarm were seen. On November 14, 1898, however, larger numbers appeared. For instance at Harvard no less than 800 were counted by 30 observers, and considerable numbers were seen from numerous other stations all over the world,[13] indeed there was enough of an increase over 1897 to raise high the hopes of everyone. The writer remembers most vividly the night because it was to watch for the expected shower that he spent, as a boy, his first second half observing, seeing in all 120 meteors in an interrupted watch extending from midnight to dawn. Some of the meteors, particularly toward daybreak, were quite brilliant.

The long history of the Leonids, the returns in 1866–1868 as predicted, and the spectacular features of a great meteor shower excited both in press and public most lively interest and expectation. Any misgivings such as those aroused by the paper just reviewed were generally unknown, and as lists of eclipses and other astronomical phenomena were the one class of predictions that was never known to fail, the average person as fully expected to see a great meteoric shower as he did to see the sun rise next morning. It is the writer's

[13] See practically every astronomical periodical published during 1898 and 1899 for accounts

personal opinion that the failure of the Leonids to return in 1899 was the worst blow ever suffered by astronomy in the eyes of the public, and has indirectly done immense harm to the spread of the science among our citizens. Certainly meteoric astronomy has never since in America recovered its proper place in the attention of professional astronomers as a class, and only recently in Europe have trained men of great ability again begun to turn serious attention thereto, with a very few honorable exceptions who labored during twenty years in an almost neglected field

Returning to 1899, great preparations were made at many observatories, and as the writer's experience was typical, he may be pardoned for giving it. Prof. O. Stone, then director of the McCormick Observatory, organized six parties of two observers each, which were scattered at intervals along a line 40 miles in length running north and south through the observatory itself. These parties were equipped with cameras and the usual maps, etc Mr. J. A. Lyon, with the writer as assistant, was sent to Scottsville, Va., about 20 miles south. Here the telescope and attached cameras were set up in a front yard which commanded an excellent view towards the east and well to the west of the meridian Unfortunately the sky was illuminated by a full moon, which of course cut out all chance of seeing faint meteors, but if they returned as on former maxima we knew there would be enough brighter ones to completely fill our time and give a very brilliant spectacle The night of November 14 was cloudy, and observation could only be made for short times between holes in the clouds. Only seven meteors in all were recorded. The next night was cold and brilliantly clear and observations were carried on from midnight to dawn. Only 20 meteors were seen, 14 being Leonids. The twenty-four years that have since elapsed have not been enough to lessen the memory of the bitter disappointment felt when dawn finally broke and the last hope of a shower faded away! It is true that at other stations throughout the world a few more were seen but nothing that could by the wildest stretch of the imagination be called a shower.

The real maximum came November 14, 1901, when at some stations meteors attained a considerable rate. Prof. W. H. Pickering, in an article[14] containing a résumé of the observations of this date deduced

[14] *Popular Astr*, 10, 400, 1902 Also an article by R. B Taber, in same volume, p. 403.

that in the region from the West Indies to California, and in general south of latitude 30° N., a really fine shower occurred. The hourly rates quoted were from 225 to 800 as a maximum (one observation of "countless" being discounted). In the eastern part of the United States a good shower was seen, but such large rates were not obtained R. M. Dole in Massachusetts, for instance, saw 256 meteors, 222 being Leonids, from 12^h to 18^h, but as he mapped some many others must have been lost from his total counts. The writer in Virginia, from 16^h 50^m to 18^h 18^m saw 88, some very brilliant and leaving splendid trains, one of the latter lasting over three minutes. On the previous and following nights, during 9 hours watch, the Leonid rate was only 6 per hour. Moonlight and cloud in 1902 interfered, but there was a fair return in 1903. Altogether to the present the writer has observed on from one to four nights about the maximum of the Leonids during 1898, 1899, 1900, 1901, 1903, 1904, 1906, 1907, 1909, 1911, 1912, 1915, 1917, 1919, 1922, and 1923, but in no case has anything worth mentioning been seen since 1903, yet always a few Leonids appear. This has been the general experience of others who have tried year by year to follow this most important stream.

The probable explanation that has been given for the failures of 1899 and 1900 is as follows. The main group of the Leonids takes three years to pass a given point. When the group approached Jupiter's orbit the planet was still too far away to greatly perturb the front members Hence the 1898 shower was fairly good. But when the part which should have reached us in 1899 was passing, Jupiter meantime had come very much nearer. It was about equally near on the other side of the stream when the group that should have reached us in 1900 passed But where the rear end, that reached us in 1901, passed nearest the planet the latter meantime had moved on a long way. Hence while able to completely switch out the main (1899 and 1900) groups, the planet was not able to perturb the 1898 and 1901 groups sufficiently to keep us from meeting them in fair numbers. So far as known the perturbations from 1900 to 1933 have not yet been calculated by any one, hence it is too early to hazard a guess as to whether the accumulation of perturbations may or may not again switch the main group exactly across the earth's orbit at the critical date. Yet it is only fair to point out that evidently in the past 1000 years some dates when returns might have been expected are not recorded in any annals, and it is not unscientific to believe that for a fraction of these dates at least the main stream missed the earth Such being the case its failure, or partial failure, in 1898–1901 is not to be considered final, and the future may yet permit us to see returns as rich as those recorded in 1799 and 1833, for instance. At any rate all interested in meteors will make adequate preparations in the interval 1932–1935 so as not to be caught unawares, if indeed a grand shower should appear.

Although the questions as to the past or future of the Leonids will not be considered here, still it will be helpful to state in what way we now believe they are distributed around their orbit. This orbit has a major axis of 10.34 astronomical units (which $= 9.5 \times 10^8$ miles) and its eccentricity is 0.90. It crosses the earth's orbit at an angle of 163°, the actual motion of the meteors therefore being retrograde or clockwise as seen from the north side of the earth's orbit. (All the planets move direct or counter clockwise.) This puts its aphelion point about 18.0 astronomical units from the sun, the mean distance of Uranus being 19.2. But due to their respective orbit planes being inclined about 16° to one another, the meteors could never approach this planet within 5 astronomical units. We believe that the orbit, so-called, occupies the center of a closed cylinder or tube of space, with an extreme oblique section of about four and a half million miles in diameter, which is sparsely filled with meteoric bodies, moving in orbits more or less of the same size and shape, and with major axes merely parallel and not very greatly different. That in a portion of this tube meteors are very densely packed along the axis for such a distance that they will take approximately three years to pass a given point, this interior cylinder not being more than 120,000 miles in diameter.[15] In the space all along the axis, not occupied by the main group, meteors are to be found in small and varying numbers. About the thirteenth to the sixteenth of every November the earth goes through the tube, but only when it intersects the dense, shorter interior cylinder can a great shower occur.

Incidentally it is very easy to overestimate the number of meteors found in any given unit of volume. The first trustworthy estimate noted was made for November 13, 1867, by Newcomb,[16] with regard

[15] Based on the supposition that it takes the earth 6 hours to go through it centrally.

[16] Abstract in *Am Jour Sci.*, (II), **45**, 236, 1868.

to the Leonid shower of that date. The observed rate was 3000 per hour, and it was supposed that all were seen which fell within 75 miles of the observers. The relative velocity was taken as 44 miles per second. He then calculated that there was one meteor for every 900,000 cubic miles of space. Without discussing the exactness of his data, it must be at once realized that the densest part of the richest meteor stream has plenty of space between its individual members. Even at that its cross section is so immense that Newcomb estimated one million meteors passed every second, so that the total number in such a swarm is practically incalculable.

In closing this chapter it may be remarked that the fact has long been well known to astronomers that the periods of Halley's Comet, for instance, vary between 74 and 79 years. Halley's Comet also moves in a retrograde direction, the inclination of its orbit being almost the same as that of the Leonid orbit, hence we are safer than usual in drawing conclusions by analogy. The moment therefore that the connection between a meteor swarm and a comet was proved it could have at once been expected that similar variable periods would be found for the former. We therefore do not necessarily believe that the drawing out of the main Leonid swarm is going on very rapidly just because the shower comes later or earlier by a year or two than indicated by the 33 25 year period, still the best average value to assume.

CHAPTER V

The Perseids

The Leonids as we have seen may furnish us with splendid showers at three epochs per century, but on most of the intervening years very few of them are visible. In the month of August, however, comes the maximum of another stream, whose radiant is in the constellation Perseus, and whose members are therefore called Perseids. Instead of coming in great numbers some years and then falling off to almost nothing in others, the Perseids appear with no remarkable variations in numbers practically every August. Recent researches seem to show that the stream is cut by the earth about the middle of July and is so wide that we do not leave it before the middle of August. At present the maximum may be expected on August 11. Meteors are still very frequent for the next two nights, after which there is a sharp decline, it being very difficult to determine a good radiant, due to small numbers, by August 17.

From what has been said we have less reason to expect mentions of the Perseids in very ancient writings than of the Leonids, and indeed it is to the careful Chinese records of meteoric phenomena that we owe the earliest accounts. Historically, A. Quételet in his *Catalogue des Principales Apparitions d'Étoiles Filantes*, which appeared in 1839, called wide attention to the fact that every August a large number of meteors, coming from the region of Camelopard, were to be observed about the ninth of the month This fact he had actually first announced in 1836. He states that he was aided not only by his own observations but by the following records that he came upon. First in a work by Musschenbroek, 1762, appeared the following statement: "Stellae (cadentae) potissimum mense Augusto post praegressum aestum trajici observantur, saltem ita in Belgio, Leydae et Ultrajecti." Second in a book by Dr. Foster, 1827, *The pocket encyclopedia of natural phenomena, etc*, in which he copies from the manuscript of a monk of the previous century the statement that meteors were frequent on August 10. The same writer quoted a tradition that among the Irish peasants the meteors were known as the "tears of St. Lawrence," whose festival happened to come on the tenth. Quételet makes the statement that despite the few documents available, for nearly every year during the past twenty-five he found some statement of the great frequency of meteors on the dates in question. His own observations fully corroborated these references.

We may pause here to do justice to another name, almost unknown to astronomers though a physicist of reputation. In S. C. Walker's article, pages 139–140 (already quoted on page 31), may be found the following:

> I made mention of the observations of Professor Locke published in 1834, in the *Cincinnati Daily Gazette*. Those of the 8 and 10 of August in the same year show that, although the periodicity of the August meteors was first discovered by Quetelet in 1836 the position of their radiant and the convergent points was first discovered and pointed out by an American in 1834.

Some of the words of Prof. John Locke are now quoted from the article:

> I was surprised to discover that most of these meteors had such apparent motions as would be produced by bodies moving parallel to each other in straight lines. That is, they describe parts of great circles, which, if produced would all meet and cut each other in two opposite points . . . . By tracing the tracks of the above observations on the globe, the radiating point or pole was found near the star Algol, in the constellation Perseus and the opposite or convergent point, in the constellation Lupus. This was the course of most of the meteors . . the poles did not appear to move with the earth, but they retained their places among the fixed stars . . .

He states that in two hours (9:00 to 11:00?), he saw thirty meteors on August 8, being able to see only those in one-fourth of the heavens. The above quotation is based upon those thirty meteors. He further asked in the newspaper that others in nearby towns join him in making corresponding observations. The proof is thus clearly given that the radiation of the Perseids was first discovered in 1834, in America, as that of the Leonids had been during the previous November. On August 9, 1837, the radiant was fixed by G. C. Schaeffer[1] of New York as being about $\alpha = 55°$, $\delta = +60°$. Apparently he was quite unaware of its having previously been determined.

Late in 1837, E. C. Herrick of New Haven published a series of most important papers[2] upon the subject, at first unaware that

[1] *Am. Jour. Sci.*, (I), **33**, 133, 1838

[2] *Am. Jour. Sci.*, (I), **33**, 176 and 354 and 401, 1838.

Quételet had been carrying on a similar research in Europe. These papers contain not only all the very numerous American observations of which he could secure possession, but also notes as to all older accounts he could find in European annals. He concluded that the shower lasts at least three, but possibly ten or fifteen days, reaching its maximum on August 9. He states that the radiant is further north than in the November shower. "On this point, however, nothing positive can be stated without observations continued during the whole night." This in spite of the fact that he quotes G. C. Schaeffer's observations given above. In fairness to Schaeffer we must say that the latter observed from 8 to 15 hours, seeing from 200 to 300 meteors, hence Herrick's uncertainty seems to be caused by the rest of the observers not fixing any certain radiant point. He was wholly unaware, evidently, of the work of Locke, already quoted. In passing we may say that it is a matter of pardonable pride that in America so much of the early important work in meteoric astronomy was done, while it should be equally a matter of serious regret that at present the whole subject is neglected in the most unaccountable manner. However, as stated before, in Europe many men of ability are now turning their attention to the subject, and new advances may be expected in the near future.

Quételet, Herrick and others gave lists of many meteoric showers that old accounts reported as appearing late in July or early in August. Littrow in 1841 combined the showers recorded for A D. July 20 to 25, 830–841 and that of July 27, 1451, with that of August 9, and found for the period just one siderial year. Also Biot in 1843 ascribed to the Perseids the many ancient accounts of meteoric showers in late July. Newton in a paper already quoted (p. 29) gave a list of such showers reduced to the same date of 1850. We copy this list in full in table on following page.

The first ten dates which Newton kept together he rightly considered as all belonging to the Perseids. The last five, taken from a supplementary table, he does not pass an opinion upon. As it is now well known that a shower of Aquarid meteors has its maximum about July 28 to 30, it seems not unreasonable, in the writer's opinion, to assign the showers of A.D. 784 and 714 to that group, rather than to the Perseids. The other three cases cited can hardly be assigned to either the Aquarids or the Perseids, and must be simply considered as manifestation of showers of which we have no other certain record.

The present date of the maximum for the Perseids is generally August 11, which reduced back to 1850 would give August 10, so we see that for 1100 years the date of maximum has scarcely changed at all. This is to be explained by the orbit of the Perseids being so nearly perpendicular to that of the earth's i.e., $\iota = 114°$. Hence perturbations are less likely to shift the node than if ι were small. No certain period has been fixed for this stream, though Schiaparelli inclined to one of 108 years. There were however well marked maxima in 1909 and 1922.[3] As stated on page 42 the duration of the Perseid shower is considered to be at least a month. It has now been observed practically every year for the past century, and for eighty

A.D.	830	July 26	corresponding to A D	1850 Aug 9 2	Biot
	833	July 27		1850 Aug 10 4	Biot
	835	July 26		1850 Aug 8 9	Biot
	841	July 25		1850 Aug 8 4 .	Biot
	924	July 26-28		1850 Aug 8 1-10 1	Biot
	925	July 27-28		1850 Aug 8 8- 9 8	Biot
	926	July 27		1850 Aug 8 6	Biot
	933	July 25-30		1850 Aug 5 8-10 8	Biot
	1243	Aug 2		1850 Aug 10 6	Herrick
	1451	Aug 5		1850 Aug 10 0	Biot
	36	June 25	corresponding to A D	1850 July 20.8 .	Biot
	784	July 14		1850 July 29 0	Biot
	1022	June 28-30		1850 July 9 3 11 3	Chasles
	714	July 19		1850 Aug 2 9	Biot
	865	Aug 5		1850 Aug 19 3	Biot

or more years with reference to its radiant. Schiaparelli stated in a note (*Sternschnuppen*, p. 97–8) that a very great number of parasitical radiants are, about August 10, found quite near the main radiant of the Perseids. In fact meteor lists are packed with all sorts of minor radiants, which are supposed at that date to be near the chief one The writer believes it to be a fact, however, that with increasing experience observers find that more of the meteors seen really converge from the central point or small area. Hence while quite certain that many minor radiants do exist in the neighborhood, still he believes the larger part published are merely due to errors of

[3] Denning has just published (*Monthly No.*, **84**, 43, 1923) a paper in which he derives a period between Perseid maxima of 11 72 years

observation, which, while only inconveniently large for those persons of greatest experience, grow to more harmful size for observers who only work very occasionally on meteors. His own experience in observing the Perseids, which goes from 1899 to 1923 inclusive, and containing meteors observed on about 70 nights, in 20 different years, is the partial basis for the opinion just expressed.

A. C. Twining in 1861 published[4] his observations of the Perseids made in 1858, 1859, 1860 and 1861. His article is accompanied by a map showing the observed positions of the radiant His map and statements clearly prove that in 1861 he detected and announced the motion of the Perseid radiant from the period August 10 to 13 inclusive. While an inspection of his map shows that, doubtless due to poor projection of the charts used, his results are not so accordant as more modern ones, yet the motion is unmistakable and the writer believes that to him should go the credit for the first certain proof of such a motion. Nevertheless, to W. F. Denning of Bristol, England, meteoric astronomy is under obligation because, among many other things, he first proved beyond a doubt that, beginning in July sharply defined radiants could be observed, which night by night moved in such a way that, following a regular curve, by August 10 the position coincided with the main Perseid radiant He then found the radiant continued to move toward increasing right ascension until about August 20, when the shower ceased. He states that he first noticed this motion in August, 1877 He published his first ephemeris, based upon his own work and some of that done by others in 1899, *Memoirs Royal Astronomical Society*, 53 A revised ephemeris by him appeared in *Monthly Notices, Royal Astronomical Society*, 1901, 62, 161 (see also page 232). In the writer's opinion this proof of the motion of the Perseid radiant forms one of the most important of the advances ever made in observational meteoric astronomy. He is further glad to say that his own observations[5] have partly corroborated those of Denning upon this point during the interval July 21 to August 14—a remark not entirely valueless as the work of some European observers had shown little or no evidence of regular motion. Nevertheless he is positive that a number of neighboring radiants are in simultaneous activity, whose meteors to all appearances differ little if any from the Perseids. We are thus

[4] *Am Jour. Sci.*, (II), **32**, **444**, 1861

[5] *M* 1, *M* 2, and *M* 3.

forced to conclude that the Perseids form a very wide stream, and it is impossible to expect exact similarity in the elements of the orbits of members, which are observed at considerable intervals of time. We could only expect the inclination to be nearly the same for all the orbits. The whole question will be more fully discussed later (page 231), where the proof of the statement can be given.

As for the Leonids, so also for the Perseids, the astronomers from 1834 on for many years seem to have generally believed in short periods for such streams The following is quoted from Schulhof,[6] the original memoir not being available.

> Boguslawski, the first who undertook such a calculation (i e , an orbit for the Perseids) for the meteors observed on August 10, 1838, at two stations, did not fall into this error Considering that from new indications, the movement of the meteors took place in very elongated orbits, he regarded as justified the introduction into his calculation of the parabolic velocity, while admitting that the supposition is not sufficiently exact in all the cases of periodic meteors We find in his memoir (*Corr Ast Quételet*, t xl, 446) the judicious remark that the angle formed by the radius vector and the tangent of the orbit (angle on which depends all the elements except the period) is not too much modified by a slightly inexact supposition as to the velocity The extremely simple method suggested to him by Olbers required only the knowledge of the points of appearance and disappearance of the meteor, and the velocity reduced from the duration of the phenomenon He determined in this manner the orbits of six meteors, but feeling the insufficiency of the results deduced from such precarious data as those of the observed durations, he repeated the calculation for five meteors, adopting arbitrarily the parabolic velocity. The elements thus found have only a vague resemblance with the swarm.

Just why the latter statement was true is not explained.

A very important paper[7] on the orbit of the Perseids was published by Prof. A. Erman of Berlin in 1839. In this paper he discusses some general questions about the observations and then gives the radiant points deduced for August 9, 10, 11. These are respectively August 9, $12^h.4$ Berlin M. T. $\alpha = 44.86°$, $= +50.18°$; August 10, $10^h.6$, $\alpha = 43.88°$, $\delta = +52.39°$; and August 11, $11^h.2$, $\alpha = 48.45°$, $\delta = +51.05°$ which were based on 50, 48 and 43 observations respectively. From the size of the probable errors given we conclude that every given meteor whose projected path came within 7° at

[6] *Sur les Étoiles Filantes, Bul. Astr.*, **11**, 126, 1894.
[7] *Astr. Nach.*, **17**, 3, 1839.

least of the radiant was considered a Perseid and hence very large radiant areas were found. Nevertheless they are of great interest as indicating a real shift from August 9 to 11 in right ascension. Erman having other matters in mind seems to have totally overlooked this, which if followed up might have given him the credit of the discovery made by Twining in 1861 and by Denning in 1877. He then developed a somewhat complicated method for calculating the orbit It not being our purpose to describe at length mathematical methods, wholly obsolete, we will only say that assuming his radiant as $\alpha = 44°$, $\delta = +50°$ and successively using five possible values for the velocity, from 0.5572 to 1.42368 respectively, he calculated five orbits. The last value was for parabolic velocity, or infinite period. The first four all gave the period less than a year, but only the first gave a direct motion, with $\iota = 56.3°$ All the other suppositions gave $\iota > 100°$, or retrograde motion, the last being 123.8°. The only reason that his parabolic elements did not come out almost exactly correct seems to be that his observed declination was about 5° too far south (that is judging by observations of the radiant made from 1863 on to the present) In all cases he found, however, that the angle made with the earth's orbit was very great. He pointed out that corresponding observations of the individual meteors might give a satisfactory velocity for the stream, but never appears to have carried out his idea. His method included terms for the diurnal aberration, an advance in exactness, but not the more important perturbations due to the earth. The latter part of his article contains a discussion about whether certain cold spells could be produced by the passage of the Perseid and Leonid streams between the earth and the sun, which he believed took place about February 7 and May 12.

Almost at once (December, 1840), Prof. Benjamin Pierce of Harvard wrote a criticism[8] of Erman's paper in which he pointed out the vicious effects of neglecting the acceleration produced when the meteor came near the earth due to the latter's mass, and also an error in Erman's equations. Taking Erman's data he then argued that as a mean difference of direction of 10° was found, only the attraction of the earth could produce this. By calculating this latter for the maximum case, a meteor just grazing the earth, he

[8] *Trans Am Philos. Soc*, *N S*, **8**, 83, 1843.

found only a value of 2.6°, quite contrary to the majority of the observations. His further argument lead him to the conclusion (utterly erroneous) that the meteors move almost in the plane of the earth's orbit etc. Nevertheless the paper is of value because it points out Erman's omission of the earth's effect.

In the paper by S. C. Walker, already quoted (page 31) the orbit of the Perseids was calculated. But using the mean of the velocities as observed by Brandes, Quételet and Twining, he assumed the value for the relative velocity of only 1 112 This gave a period of 0 562 and an inclination of 78 9°, which meant direct motion. The radiant used by him for his calculation was 1840, August 9.456, $\alpha = 36\ 1°$, $\delta = +\ 55.8°$. In other words his right ascension was worse and his declination better than those used by Erman and strange to say by about equal amounts. If Walker's actual orbits were wrong, not so some of his opinions. Among the latter we first find clearly stated the possibility of meteors being the basis of the "resisting medium" necessary to explain the acceleration of Encke's Comet, and a reason given why Halley's Comet and others need not necessarily show the same effects (loc. cit., p. 112). Comparing Walker's method with that of Erman we find that the former's equations were more general, his coordinates better and above all that the equations were equally applicable to all forms of conic sections. A further quotation will be made from this article when we come to the subject of telescopic meteors (p. 151).

A. C. Twining in 1862 published an article[9] containing a very good geometrical discussion of the general effects produced upon the meteor ring of the Perseids by the yearly passage of the earth through it. He concluded: (1) That the position of the node would not be shifted by more than one or two degrees in half a million years . . . . (3) That there is an appreciable change of radiant positions relative to locality . . . . whose maximum may be $3\frac{3}{4}°$ between extremes. (4) That the terrestrial disturbance is enough to affect the perihelion distances of individual meteors many millions of miles and to expand the ring to a corresponding breadth at the ascending nodes. (5) That terrestrial disturbances do not appear sufficient to draw meteors off into permanently erratic orbits. His conclusions deserve the more respect as they were based upon a supposition of the meteors'

[9] *Am Jour Sci.*, (II), **33**, 244, 1862.

relative velocity of 26 6 miles/second, which he says he based upon results of Marsh, Herrick, Newton and himself (loc. cit., p. 249). He hazards a guess finally that the impact of a meteor stream upon the head of a comet might be *seen* while any attractive influence of one upon the other would be hopeless to detect.

To follow the logical order we must here quote part of an article[10] by Daniel Kirkwood of Washington and Jefferson College. It appeared in the *Danville Quarterly Review* for December, 1861, and was written by him in July of that year. Having in a masterly way collected and discussed the evidence available to him upon the division of comets, six cases of which he quotes, he ends his argument with the following lines:

> In view of these facts it seems highly probable, if not absolutely certain, that the process of division has taken place in several instances besides that of Biela's Comet May not the force, whatever it is that has produced one separation, again divide the parts? And may not this action continue until the fragments become invisible? According to the theory now generally received, the periodic phenomena of shooting stars are produced by the intersection of the orbits of such nebulous bodies with the earth's annual path Now there is reason to believe that these meteoric rings are very elliptical, and in this respect wholly dissimilar to the rings of primative vapor which, according to the nebular hypothesis, were successively abandoned at the solar equator, in other words that the matter of which they are composed moves in cometary rather than planetary orbits May not our periodic meteors be the debris of ancient but now disintegrated comets, whose matter has become distributed around the orbits?

In this quotation we see the first sound argument, based upon philosophical grounds, which was given to prove the connection between comets and meteors.

To the present point we have seen that the real stumbling block had been that nobody was able to get a good value for the observed velocities. Two eminent men solved the problem indirectly, but in nearly the same manner and by using the same data. The one was H. A. Newton in 1864–1865, the other G. Schiaparelli in 1866, each without knowledge of the other's work. We shall follow here the arguments as developed by both in order, as these two researches lie at the basis of all further advances.

[10] *Meteoric Astronomy, Appendix A*, by Kirkwood, 1867.

COMET 1908 c (MOREHOUSE) ON OCTOBER 1, 1908, AT 13h 43m C S T.
Exposure 2h 0m, by E E. Barnard at Yerkes Observatory

meteors, and V_0 is the velocity acquired by a particle falling from rest at infinity under the gravitation of the earth alone then instead of this last expression we must write $\pi R^2 nV\left(1 + \frac{V_o^2}{V^2}\right)$. If the earth is in uniform motion then V must represent the relative velocity of the meteors and the earth. Extending this to an indefinite system of small bodies, we may then write $N' = \pi R^2 \left(\Sigma n' V' + V_o^2 \Sigma \frac{n'}{V'}\right)$. But $\Sigma n' V' = n V_a'$, where V_a is the mean value of $V' + V'' + \ldots$, and n the mean value of $n' + n'' + \ldots$ For the other term we may also write $V_o^2 \Sigma \frac{n'}{V'} = \frac{nV_o^2}{V}(1 + \theta)$ where θ is always a positive number. If the values V′, V″ do . . not vary widely, θ is small. We then finally have $N' = \frac{\pi n R^2}{V}(V_a^2 + V_o^2 + \theta V_o^2)$, N′ expressing the number of meteors the earth meets per unit of time. But if m is the average number visible at one place we have found N = 10460 m for the whole earth. Earth's volume is $\frac{4}{3}\pi R^3$. Let M be the number of meteors in an equal volume of space, then $M = \frac{4}{3}\pi n R^3$, and $10460\ m = \frac{3}{4}\frac{M}{V_a R}(V_a^2 + V_a^2 + \theta V_o^2)$. If the hourly number is 30, then $m = \frac{30}{60 \times 60} = \frac{1}{120}$ and $M = \frac{4}{3}\frac{10460\ m\ RV_a}{V_a^2 + V_o^2 + \theta V_o^2}$ $= 116.2 \frac{R\ V_a}{V_a^2 + V_o^2 (1 + \theta)}$

Newton next calculated the average angular length of path to be 12 6°. The following section deals with telescopic meteors, from which he deduced that more than 400,000,000 of such objects also enter our atmosphere daily. This is discussed in Chapter XIII. To determine the mean distance of the meteor's path from the observer only inadequate data were at hand. He made two assumptions: (1) that those paths along the oblique line that become invisible are always the more distant ones; (2) that the paths which become invisible are distributed along the oblique line in proportion to the numbers along the line. The first gives 140 km., the second 232 km. Newton was sure the truth lay between these limits. He concluded that the foreshortening of the paths could be found by the expression $12.6° \times \frac{4}{\pi} = 16.04°$. Using this the mean length would be either 65 km. or 39 km., according to whether 232 km. or 140 km. was used for the mean distance. His opinion was that 39 km. is much nearer the truth than the greater value. The mean observed duration of 867 flights was 0.45^s, and most observers agreed that they were not over 0 5^s Taking this last as a fair average then the nearer velocities turned 78 or 130 km /sec. As either of these vastly surpasses the parabolic velocity at the earth's distance from the sun, such meteors could not then move in closed orbits. Three possible conclusions occurred to him, (1) that the length of track is too long; (2) that the observed duration is too small; (3) that very many of the meteors move in hyperbolic orbits about the sun.

In the same volume from which this is quoted, page 370, Newton gives the later estimates of the durations observed by Schmidt at Athens. The latter had records of 1357 meteors as to durations and found the mean to be 0 925^s. (It is, however, quite obvious by noting the data given by Schmidt, which were copied by Newton, that only 1357 of the 16000 meteors were so observed, and the colors were also recorded for these. As the colors of fainter meteors can never be seen, and those of very swiftly moving meteors are very hard to detect, we have every reason to believe those observed by Schmidt were the exceptionally long durations, not the fair average for the 16,000. Also very certainly those were the brightest seen, which on an average do last longer than the fainter ones.) To help out the first case Newton thought that the altitudes > 150 km. should be rejected, 57 in all. In the second that perhaps the mind could not make proper allowances for the time that elapsed, after seeing the meteor, before the eye is directed to the place of the path. And lastly he knew that the Leonids and Perseids moved in closed orbits and could not quite equal the parabolic velocity as a limit.

He finally concluded that the sporadic meteors cannot all belong to one narrow ring which has a diameter nearly equal to the earth's orbit. Secondly that a large portion of the meteors must, when they meet the earth, have absolute velocities greater than the earth's velocity in its orbit, or else that the sporadic meteors have a series of radiants at some distance from the ecliptic, and hence come from a series of rings considerably inclined to the earth's orbit. He next suggested three possible distributions of meteor orbits in the solar system. (1) They may form a number of rings. (2) They may form a disk in or near the plane of the orbits of the planets. (3)

They may be distributed at random like the orbits of the comets. Omitting the discussion of the first two (untenable) suppositions, Newton derived the formula[12] $N_o = \frac{n}{2}\left(1 + \frac{V}{V'} \cos z\right)$ where V and V′ represent the mean velocity of the earth and that of the meteors, n and N_o the number of meteors which would come from all the sphere and those which one would actually perceive in the visible hemisphere, and Z the zenith distance of the apex.[13] He calculated $N_o \div n$ for the latitudes of New Haven and Paris on the assumptions that $V' = V\sqrt{2}$, $V' = V$ and $V' = \frac{1}{6} V$. Obviously the first gave parabolic orbits. The derived numbers were then compared to the tables of Herrick and Coulvier-Gravier for the rates of different hours of the night. He concluded that velocities greater than parabolic are indicated, but since the data are uncertain and the reasoning contains certain assumptions, all he felt certain of was that the orbits cannot be even approximately circular, but that meteor orbits rather resemble those of comets

Lastly assuming $V' = \sqrt{2}$ he calculates that at any given moment there would be 13,000 meteors, visible to the naked eye, in a volume of space equal to that of the earth and at the earth's distance from the sun. He further thought that the space near the earth's orbit was not more densely filled by meteors than other parts of the solar system. And further that these meteors could not be regarded as fragments of former worlds, but rather the material from which worlds are forming.

The reader will perceive that by now many weighty reasons had been adduced for the belief in the connection of comets and meteors. It only remained for the proof itself to be given in which it could be shown in specific cases that a meteor stream and a comet follow the same orbit in space. But before giving this certain opinions, as yet not quoted, should be mentioned so that the complete argument, as Schiaparelli found it, may be clear to the reader. In this brief résumé Schulhof's memoir, already referred to on page 47, is very closely followed. The views of Halley, Chladni, Olmsted and many others on this point, having already been quoted, we will not here

[12] This formula had already been derived by Herschel and was soon after independently derived by Schiaparelli.

[13] Note that $\cos z = \cos\phi \cos\theta$.

repeat them Kaemte and Littrow accepted the views of Chladni. The Abbe Raillard in 1839 found analogies between meteors and comets. Capocci believed aurorae, meteors, meteorites and comets had a common origin, that they are conglomerations of cosmic atoms united by magnetic attraction and that comets are only large meteors. In 1841, S. M. Drach,[14] noting the great velocity and various inclinations of their orbits was inclined to assign to them a cometary origin. Truth forces us to say that part of his data and some of his other conclusions were wholly wrong. Foster in 1843 affirmed that whenever a great comet appeared he was certain to see very many meteors. Finally Reichenbach[15] in 1859 published a memoir upon the relation of comets to meteorites He considered comets as aggregations of primitive matter of which the molecules have the tendency to concentrate more and more and that meteorites come from the condensation of comets. Finally with regard to the excellent deductions of Kirkwood, quoted on page 50, which were quite the most logical of all, fairness to Schiaparelli compels the statement that not until the publication of Kirkwood's complete book in 1867 did his views reach European astronomers, and meanwhile were wholly unknown abroad.

At the risk of some repetition, the argument leading up to his final proof will now be given as developed by Schiaparelli, *Sternschnuppen*, page 46 et seq.

The phenomenon of daily variation of meteors was shown by the work of Herrick and Coulvier-Gravier, and when it was found that it might be largely explained as a consequence of the motion of the earth in its orbit combined with the motion of the meteors, then it could be proved that the law of daily variation would give information as to the absolute velocity of meteors in space. The conclusion was that it must be greater than that of the earth. The basis for the above lay in these facts. because the researches of Erman and Newton had shown that for the Perseids and Leonids the orbits had a great inclination and retrograde motion respectively, then it was not probable that the meteors belonged to the same class of bodies as planets. Actual observations of velocities, by many observers, gave grounds for the belief that meteors passed the earth's orbit at all angles of inclination. The assumption that meteors met the earth from all

[14] *Monthly Not., R A S*, **5**, 126, 1839-43

[15] *Poggendorf Annalen*, **106**, 438.

directions was not opposed to common sense and could be proved by observation.

If we assume the earth to be stationary or merely rotating upon its axis in space, in the midst of a cloud of projectiles, obviously every part of its surface will be struck equally often. We could hence have no hourly variation of meteors. Again if the earth moves forward with a tremendous velocity (such as that of light), incomparably greater than that of the meteors, all hits will be on the hemisphere which is on the forward side and the earth will leave a wholly empty space behind it. Under this hypothesis we will be able to observe meteors only so long as the point toward which the earth is directed (in its motion around the sun) lies above our horizon. Once this point has set at any given place no more meteors could be seen there until it again rose, approximately half a day later. Finally the logical assumption is that earth and meteors move with velocities of the same order of magnitude. In this case the hourly number would vary according to the height of this point above the horizon. This point is called the meteoric apex, or more briefly from here on, the apex. Obviously it always lies in the ecliptic and is defined at any moment by the direction of the tangent to the earth's orbit at the point the earth occupies at that moment The orbit not being a perfect circle, this point is not exactly 90° of longitude less than the sun's position but the eccentricity of the earth's orbit being small, it only varies between the extreme limits of 89° and 91°. Obviously when the apex is in the zenith, the greatest number of meteors can be seen, according to this hypothesis. On an average the apex will be on the meridian at upper culmination about 6:00 a.m. or 18 hours astronomical time, and it will have its lower culminition at 6:00 p.m. or 6 hours. This at once shows that more meteors per hour are to be expected after midnight compared to the hourly rate before. What has been found by this reasoning coincides with the laws deduced from the observations of Coulvier-Gravier and Schmidt. The mathematical development of Schiaparelli will be given in Chapter XV.

Schiaparelli stated that it was not difficult to formulate such a law, based upon Coulvier-Gravier's observations, when he assumed that all meteors had a certain mean absolute velocity when they entered the earth's sphere of attraction, in which the influence of the earth's attraction and the atmospheric retardations were neglected. As a result it was found that the mean absolute meteoric velocity was 1.45 times that of the earth, for convenience assumed to be 1.41 = $\sqrt{2}$, which would exactly correspond to the parabolic velocity. This research, published in September, 1866, he stated brought him to nearly the same conclusion as Newton had come to (see page 54), which he added was the more remarkable as he had no knowledge of the work of the former.

At once he saw that approximately parabolic orbits may be in many cases long ellipses, and that in either case such orbits resemble those of comets; also what was known about the Perseid and Leonid orbits lent support to this resemblance. Halley long before had given out the theory that both comets and meteors come from the region of the stars. As meteors come to us already formed into systems the conception is reasonable that they were formed into systems in the depths of space and are accumulations of tenuous matter. If we inquire what will happen to such an accumulation on nearing our system we find that for each such tenuous cloud, attracted by the sun, whether it consists of continuous matter or is divided into little parts, the law of attraction must of necessity draw out this incohesive mass into a long and tenuous stream, bent into a curve, which in the parts near us barely differs from a parabola and generally is a very long conic section. Schiaparelli stated that he believed this to be the key to the formation of the non-periodic meteor streams.

For a known periodic stream as the Leonids, he assumed that a cosmic cloud, which is not yet drawn out into a stream and which is therefore relatively denser than those just mentioned, can on its approach to a large planet be changed into an orbit of short period and shorter perihelion distance. On its passage through perihelion the differential attraction of the sun upon its parts can draw it out, and the cloud will more and more approximate to a stream, which in the end will close itself and form an elliptical ring.

These hypotheses fitted in well with the known facts and the cosmic hypotheses of Sir William Herschel and La Place. For was it not very probable that matter, in every degree of tenuity, was scattered through space? Further the shapes and various inclination of meteor orbits resembled those of comets. And finally M. Hoek[16] had shown shortly before, with considerable probability, that

[16] *Astr. Nach.*, **45**, 49, 1857.

certain series of comets came from space, not as individuals only, but as members of "families," thus giving another analogy. Lastly various appearances had been reported, from time to time, which seemed to lie between comets and meteors. All these combined reasons made him attempt to calculate the parabolic orbit of the Perseids and Leonids. Basing his calculations for the former upon the coordinates of the radiant point found by A. S. Herschel[17] and derived in 1863, and assuming that the radiant was at $\alpha = 44°$, $\delta = +56°$, and that in 1866 the maximum came on August 10.75, he calculated the following orbit, in which the elliptical motion of

	PERSEIDS, 1866	COMET 1862 III
Perihelion passage	July 23 62	1862 Aug 22 9
Node .	Aug 10 75	
Longitude of perihelion	343° 48′	344° 41′
Longitude of node . . .	138 16	137 27
Inclination	115 57	113 34
Perihelion distance . . .	0 9643	0 9626
Period . . .	108 years?	121 5

the earth was allowed for but not the almost negligible effects of diurnal aberration.

For this calculation he used a method analogous to that of Erman (see page 47) in which there are two possible solutions, one giving a negative solution corresponding to the point in the sky opposite the observed radiant. The relative velocity of the meteors came out about 60 km./sec., these observed by A. S. Herschel and from which the radiant used was derived gave a value of about 56 km./sec., an excellent accord for such work. The attraction of the earth would vary the result only a little. Comparing his elements of the Perseids with those of the comet given by Oppolzer,[18] which are placed above side by side for comparison, he concluded that the two systems of elements differed from each other only by a value which could easily be explained by the lack of exactness with which the node and radiant of the Perseids could be fixed. The assumed period of 108? years rested upon the following. In the catalogues of Quételet and Biot

[17] *Proc. British Meteorological Soc.*, **2**, 19, 1863.
[18] *Astr Nach*, **69**, 87, 1867.

he found showers mentioned which were of about the right date, in the following years, the table being copied as given:

830, 833, 835, 841 A D.
925, 926, 933,
1029
1243
1451
1779, 1784, 1789

These data Schiaparelli believed indicated a period of about 108 years, but with a maximum extending 20 or 30 years, while that of the Leonids extended about 3 years only. All of the above calculations were finished in November, 1866, and published in December of that year

Both Newton and Schiaparelli deserve great praise for their valuable work, which to a certain point brought them to almost the same result, Newton, however, stopped half way to the goal. To Schiaparelli, therefore, goes justly the undivided glory of first proving the connection of a meteor stream with a comet, for it was his genius that bridged the gap which had so long stood in the way of any advance such as he made The only reason that he did not at the same time discover the connection of the Leonids and Temple's Comet seems to have been that, while his orbit of the Leonids was indeed ready, the orbit of the comet only appeared in January, 1867. This gave C. F. W. Peters the chance to anticipate him by four days, because Peters edited the *Astronomische Nachrichten*, in which the comet's orbit appeared and of course had the manuscript, while Schiaparelli, being in Italy, had to wait for his printed copy to arrive.

Before leaving the Perseids it may be said that of all the annual showers they are most certain to return with average richness. Also as they come in August, when the nights are still comfortably warm, they furnish the most excellent opportunity for a person casually interested in meteors to see a good shower. While quite numerous from August 5 to 15, yet from August 10 to 13 one is certain to see large numbers, particularly after midnight. Of all epochs of the year this is the best time for a would-be observer of meteors to start his labors. In the United States the radiant rises in the northeast, shortly after dark, and is not far from the zenith at dawn. Many of the Perseids leave very bright and enduring trains which, being beautiful phenomena, add to the interest in observing these meteors.

CHAPTER VI

The Lyrids

The discovering of two meteor streams moving in practically the same orbits as comets turned the interest of astronomers to the subject, as meteors had suddenly grown more worthy of attention in the opinion of many. It therefore followed that in February, 1867, Prof Edmond Weiss of Vienna[1] had the happy idea of calculating on what dates in the year the earth passed nearest to the orbits of those comets for which orbits were available. Those with elliptical orbits were particularly studied He thus found comet 1861 I, which at the descending node comes within less than 0 002 astronomical units of the earth's orbit on April 20, and Biela's Comet which at the descending node comes within 0.018 astronomical units of the earth's orbit about November 28. He also found in published records numerous accounts of meteors that appeared about April 20, the Lyrids, and at the end of November or in the first days of December, the Andromedids or Bielids. To this extent Weiss has the credit of the two discoveries.

Previously, in 1861, C F. Pape[2] had calculated the comet's orbit, and made the remark that according to his figures on May 12 it came within 0.0012 of the earth's orbit. He made no statement about any meteors possibly resulting therefrom. Oppolzer that same year calculated the definitive elements of the comet, giving the distance mentioned as 0 0022 in longitude 30°, which corresponds to where the earth is found about April 20. J. G. Galle of Breslau made the actual calculations[3] which proved the connection For this purpose he assumed an elliptical orbit for the Lyrids, whose semi-major axis was the same length as that of Comet 1861 I. These elements follow in the first column, while in the last three are given parabolic orbits, calculated by S J. Corrigan,[4] at Washington, for radiants determined on three different dates in 1886.

[1] *Astr Nach*, **68**, 382, 1867
[2] *Astr. Nach*, **55**, 206, 1861
[3] *Astr. Nach.*, **69**, 33, 1867.
[4] *Sidereal Messenger*, **5**, 146, 1886.

RADIANT DATE	277° 5, + 34° 6 APRIL 20	COMET 1861 I	260° 0, + 33° 5 APRIL 18	267° 0, + 33° 0 APRIL 19	274° 0, + 33° 5 APRIL 20
π	236°	243° 42′	255° 42′	248° 54′	240° 34′
Ω	30	30 16	29 05	30 04	31 03
ι	89	79 46	71 21	77 29	81 29
q	0 955	0 9270	0 8478	0 8944	0 9270
log a	1 746	1 746			
e	0 9829	0 9835			
Rel. vel.		30 m/s	28 m/s	29 m/s	30 m/s

The first radiant is that from the 1864 *Report of the B. A. A.* by Prof. A. S. Herschel, the last three were observed by W. F. Denning[5] and depend upon 6, 10 and 14 meteors respectively. Denning stated that his radiants were very sharp. Galle mentioned the wide differences between the position used by him and many others formerly derived. He therefore felt some uncertainty, from this cause, as to his deductions. In his valuable paper he called attention to the series of former great showers which might have been Lyrids, copying from Newton's paper on the subject. We again quote from the latter these dates reduced to the epoch of 1850.

B C.	687 March 16	corresponding to A D 1850	April 19 9 . .	Biot
	15 March 25		April 19 6	Biot
A.D.	582 March 31		April 18 1	Chasles
	1093 April 9 6		April 20 7	Chasles
	1094 April 10		April 20 8	Chasles
	1095 April 9 6		April 20 2 . .	Herrick
	1096 April 10		April 21.3 .	Herrick
	1122 April 10 6		April 20 2	Herrick
	1123 April 11		April 20.4	Chasles
	1803 April 19 6		April 19 9	Herrick

Compare with these the following showers for which the exact date is not known.

A D	590 before Apr 4 ??	corresponding to A D 1850	Apr 22 1	Chasles
	741 before Apr 13 ?		Apr 28 3	Chasles

[5] *Siderial Messenger*, **5**, 106, 1886.

From this table we see that the history of the Lyrids goes back possibly for 2500 years, and they are still appearing though in very small yearly numbers. Newton gives the following additional list of ancient showers in April:

A.D	401	April 9	corresponding to A D 1850	April 29 2	Biot
	538	April 6		April 24 4 .	Chasles
	839	March 29		April 12.2	Chasles
	839	April 17		April 30 9	Biot
	840	April 1		April 15 9 ..	Chasles
	927	April 17		April 29 3 ..	Biot
	934	April 18		April 30 8 .	Biot
	1000	April 4		April 15 9 ..	Chasles
	1008	April 2		April 13.6 .	Biot
	1009	April 16		April 27.6 . ..	Chasles

The excellent accord of the ten dates in the first table makes it very improbable, remembering also the high inclination of the orbit, that there could have been such changes in the node as would permit many of these last tabulated showers to be Lyrids. The only ones that seem possible would be those in A.D. 538, 840 and 1000. With regard to those at the very end of the month we shall have a few words to say later (see page 79)

As all textbooks on general astronomy naturally lay such stress upon the great showers of the Leonids and Bielids, many readers would be inclined to believe no very remarkable showers ever came from any other radiants To correct this impression, and because the accounts are little known, some of the appearances of the Lyrids will be referred to. The first possibility is from Biot's *Chinese Catalogue,* B.C. 687 (23 March). This account has already been quoted on page 2. Biot's second account is, however, clear enough that a meteoric shower is being described "March 27, 15 B.C., after the middle of the night, stars fell like a rain; they were 10° to 20° long; this phenomenon was repeated continually. Before arriving at the earth they were extinguished." No one can doubt that this was a real meteoric shower.

Omitting other accounts we will take up the last great shower of Lyrids in 1803, well seen in the eastern part of the United States from North Carolina to New Hampshire. In 1839 E. C. Herrick was able to secure and publish four newspaper accounts of this shower, which

COMET RORDAME, JULY 13, 1893
Photograph by Barnard at Lick Observatory

he had fortunately discovered. Far the best of the four will be quoted at some length, the original appearing in the *Virginia Gazette* of Richmond, April 23, 1803.

> Shooting stars This electrical phenomenon was observed on Wednesday morning last at Richmond and its vicinity, in a manner that alarmed many, and astonished every person who beheld it. From one until three in the morning, those starry meteors seemed to fall from every point in the heavens, in such numbers as to resemble a shower of sky rockets. Several of these shooting meteors were accompanied with a train of fire, that illuminated the sky for a considerable distance One, in particular, appeared to fall from the zenith, of the apparent size of a ball of 18 inches diameter, that lighted for several seconds the whole hemisphere. During the continuance of this remarkable phenomenon, a hissing noise was plainly heard, and several reports resembling the discharge of a pistol. Had the city bell not been ringing, these reports would probable have been louder [There was a fire and the people were on the streets. It is very probable that the reporter of that day merely heard some noise in the city] The sky was remarkably pure and serene, and the visible fixed stars numerous the whole night . . . This circumstance of the shooting stars descending within a short distance of the ground, is however, a fact (!) highly important to be known; as it has been generally supposed that meteors only proceed in a horizontal direction and never fly perpendicularly upwards or downwards. Those which we particularly remarked, appeared to descend in an angle of 60° with the horizon; but as the smaller ones were so numerous and crossed each other in different directions, it was only possible to ascertain with any precision the paths of the largest and most brilliant.

The latter part of the article has been quoted to show the opinion of well-informed people of the day. But these opinions are very scientific compared to the following quoted from "*Medical Repository*" (New York), 2d. Hex., Vol. 1, 1803–1804, p. 300: "The modern opinion of these appearances is, that they consist of phlogistous gas (inflammable air) catching fire in the upper regions of the atmosphere. But it is not easy to explain wherefore the air of so many parts of the continent was so overcharged with hydrogeneous vapor so early in the season." In justice to this article of which we quote the closing lines, one piece of scientific information was really included in it. An observer is stated: "to have counted 167 meteors in about 15 minutes, and could not then number them all. This display lasted from one until three o'clock in the morning."

As for the radiant point E. C. Herrick fixed its position at α = 273°, δ = + 45°, April 18, 1839. Other American observers that same year roughly confirmed this same position. It appears that

Arago, in 1835, first asked if April 22 was not a date on which meteors frequently appeared. Benzenberg made observations in 1838 and 1839, and had poor results. Herrick in 1838 and 1839 confirmed the presence of the Lyrids, but in small numbers. We can only add that for many years until 1922 the number seen per hour has usually been pitifully small for a shower, yet a persistent observer usually could obtain a radiant. On April 21, 1922, a rather strong maximum[6] occurred early in the evening, as seen in eastern Europe. Considerably more than one Lyrid per minute could be counted by one observer, and most of the meteors were fairly bright. There is no doubt that this must be considered the maximum display for half a century or more past. The period of Comet 1861 I is 415 years, but whether the Lyrids have such a period is absolutely unknown Certainly there are no accounts of showers in A.D. 1446, 1031, etc., when we might have expected that the comet had passed us. Owing to the shower's wholly unknown period, prediction of returns are impossible.

In 1914, Hoffmeister published[7] a paper on the Lyrids, basing his results upon recent observations A new method of treatment was used for the discussion of the velocities, which led to the conclusion that, if we accept the observations as correct, the meteors moved with a strongly hyperbolic velocity. However, he felt that probably the observations themselves were partly wrong, due to certain systematic errors. It can only be added that if future researches confirm the hyperbolic velocities for these meteors, one of the four classical cases of connection with a comet's orbit will have been put strongly in question. This paper should, however, be studied as an important addition to the whole subject, irrespective of the actual numerical results.

Finally it may be asked whether many meteorites fall on the dates in April when the Lyrids appear. Farrington gives (up to 1910) a table showing the number which have fallen for every day of the year. For April he gives 29 falls, with one each for April 17, 18, 19, and 20, but with none on April 21, 22, and 23. We conclude that meteorites are during the Lyrid period less numerous than in other parts of the month. One iron meteorite, however, fell in England April 20, 1876.

[6] *Pop Astr*, **31**, 172, 1923.
[7] *Astr. Nach*, **200**, 205, 1914.

CHAPTER VII

The Bielids or Andromedes[1]

The history of the meteors which are connected with Biela's Comet and of that body itself forms one of the most fascinating and important chapters in the development of meteoric astronomy. To understand the case fully it will be necessary to take up in the first place the history of the comet and follow it through its remarkable career.

In 1772 Montagne, in Limoges, France, found with his small telescope a faint comet which had a tail about 4′ long. The comet never became visible to the naked eye and, having been observed a few times at the Paris Observatory also, disappeared from view. In 1805 Pons discovered a comet, which was visible a month and could be seen by the unaided eye even in strong moonlight. Again in 1826 von Biela discovered a comet, to which his name was given. But once its orbit was computed it was seen that it was merely a return of the comets seen in 1772 and 1805. However, its last appellation held, and so its third discoverer has the honor of his name being thus perpetuated.

Briefly, its orbit is elliptical with an eccentricity of 0.76, and its period only about 6 years. When at aphelion it is near the orbit of Jupiter. As it moves with a direct motion, with the low inclination of 13°, the chances for strong perturbations by Jupiter are great. A most interesting circumstance was pointed out by observers in 1832, namely, that the nucleus of Biela's Comet would pass within 20,000 miles of the earth's orbit about December 3 of that year. This means that perturbations by the earth can also greatly change its orbit. (See also Chapter XX.)

Coming toward the sun in 1832 owing to the announcement that its orbit was so near the earth's, it produced a panic among many persons who feared that the two bodies actually would meet. The comet really passed the critical point a month before the earth reached it. In 1839 the comet when it came to perihelion was be-

[1] *Variations in spelling are found, e g., Andromedids, etc.*

hind the sun and hence not seen In 1845 the event happened which made the comet famous. On December 29 of that year Herrick and Bradley at Yale saw, beside the main comet, a small companion comet.

Changes now occurred rapidly. The faint companion grew and both comets developed tails. Then the smaller one developed two tails. Next the larger one changed the shape of its head. Its nucleus divided; two tails were formed for it alone, and an arc of light stretched from one to the other. In February the companion became brighter. Three tails were developed by the original comet, and three or more cometary fragments were seen around the nucleus. Forces were at work within the comet, violently breaking it to pieces, the nature of which was wholly unknown. The return in 1852 was eagerly awaited, but when the comets—for we can hardly use the singular any further—appeared they were both faint and over a million miles apart. There were remarkable changes in brightness, and unfortunately it was quite impossible to decide which of the comets was the original larger one, seen in 1845. This was due to the fact that when orbits were calculated for the two visible in 1852, one orbit fitted that of the 1845 return as well as the other. The comets both disappeared from view in September, 1852. In 1859 they were unfavorably placed with regard to the sun for detection, but in 1866 they were carefully searched for under favorable conditions. But from that night in September, 1852, to the present, Biela's Comet has never been seen

The first mention of meteors in connection with this comet was possibly that at St. Petersburg, December 5, 1741, a large number of meteors was seen. The next mention is that in the observations of Brandes, who travelling in Germany, December 7, 1798, saw 400 meteors in a few hours. As he presumably was in a coach, he could have seen only a small fraction of the whole. On December 7, 1830, Abbe Raillard recorded that many meteors were seen in France. On December 6, 1838, the meteors were four times as numerous as on an ordinary night. They were observed in Belgium, France and America. The radiant was roughly fixed as in Cassiopeia. This seems to have been a maximum return, compared to intermediate and later ones. On December 6, 1847, Heis, in Germany, fixed the radiant at $\alpha = 25°$, $\delta = +40°$. He later placed it at $\alpha = 21°$, $\delta = +54°$, probably on the basis of subsequent observations. Finally, one of the radiants of Zezioli for November 30, 1867, is at $\alpha = 17°$, $\delta = +48°$. It must not be understood that these later radiants were based upon large numbers of meteors, but only enough certainly to show that a meteor stream was present. It will be noted that not until 1867 did the date of maximum shift back into November.

This last date brings us to the time when Schiaparelli's work had started many others into similar research It therefore was not strange that both Weiss and d'Arrest, in 1867, and within a few days of one another, announced that the Andromeda meteors moved in the orbit of Biela's Comet. D'Arrest remarked that from 1798 to 1838 was exactly six complete revolutions of the comet, which helped to remove the possibility of a mere fortuitous similarity between the orbits of meteors and comet. On this basis he predicted another shower for December 6, 1878, which did not occur.

Shortly afterwards Weiss, in an extensive article,[2] took up the question again. Using Hubbard's elements for the three returns in question he derived the resulting radiant points for accompanying meteors:

YEAR	☊	DAY	α	δ
1772	258 4°	December 10	18 7°	+58 1
1826	251 8	December 4	22 8	+47 7
1852	245 8	November 28	23 4	+43 0

These figures show the rapid decrease in the longitude of node, and that the date of node passage therefore comes earlier and earlier. Of course the radiant must also change, with the other elements. Weiss then said that it might be possible, by observing the date of the meteor shower, to determine for the Bielids, as Adams had done for the Leonids, the secular variations due to the planets. In this way the extent of the participation of the meteors in the orbital changes of the comet might be detected. In any case the radiant based on the observations in 1838, when the corresponding orbit was calculated, fitted the 1772 orbit better than that of 1852, when the comet last appeared. He felt it probable that there were good chances for observing the phenomenon in

[2] *Astr. Nach.*, **72**, 81, 1868.

1872 or 1879, provided the two nuclei of the comet were not too far from the earth, but about November 28, not December 6, as the node in 1852 was passed in November. He said also that the very uncertain elements of Comet 1818, derived by Pogson, were quite similar to those of Biela's Comet. But Comet 1818 could not possibly be that body itself due to the date on which Comet 1818 went through perihelion. Yet it might be a fragment of that body, which had been thrown off centuries before.

A. S. Herschel, basing his article[3] mostly on that just quoted, urged all observers to be ready the last week of November and the first week in December, particularly December 4 and 7, both in 1872 and 1873 He gave a summary of the observations up to date and of Weiss's article, yet does not seem to have taken very seriously the definite prediction of that astronomer as to the earlier date of maximum. His article is of particular interest in showing the great effects of zenith attraction on a radiant, the meteors from which overtake the earth and hence have a low relative velocity.

On November 27, 1872, the prediction of Weiss was brilliantly confirmed by a splendid return Indeed Newton at New Haven, Conn., saw a good beginning of the shower as early as November 24, when 250 meteors were seen before midnight. In America, on November 27, the main display was over before dark, but in Europe very great numbers were seen. For instance, the number counted at Greenwich by one observer, from $5^h\ 30^m$ to $11^h\ 50^m$, was 10,579. Near Nottingham E. J Lowe, from $5^h\ 50^m$ to $10^h\ 30^m$, observed 14,665 At Moncalieri, Italy, P. F. Denza, from $6^h\ 0^m$ to $12^h\ 30^m$, counted 33,400 meteors, four observers participating. Many other observations of a similar nature could be quoted, but enough has been said to show the general rate of the shower. The meteors were described as being usually fainter than the Leonids of the 1866 shower.

As the shower was ending it struck Klinkerfues, who was then in Germany, that, if this was the main body of the comet which we had passed through, then the comet itself ought to be seen in exactly the opposite direction to the radiant point, as it went away from the earth He therefore telegraphed Pogson at Madras, India: "Biela touched earth November 27. Search near θ Centauri."

[3] *Monthly Not*, *R. A. S.*, **32**, 355, 1872.

Pogson found a comet in the direction indicated. He saw it on two successive mornings, both times with a decided nucleus, on the second with a tail 8′ long. Clouds and rain then came, and when he again had clear weather it was too late. This was the last seen of any comet moving in the orbit Biela's Comet once followed. Newton[4] computed that Biela's Comet itself was about 200,000,000 miles distant, hence this could hardly have been a piece of either of the parts last seen in 1852, but he considered it a fragment thrown off long before.

On November 27, 1885, P. F. Denza, from $6^h\ 0^m$ to $10^h\ 08^m$, observed 39,546 meteors, with from four to two observers. He calculated that four persons, watching the sky continuously, would have seen 62,300. Capt D. Wilson-Barber, at sea, longitude 60° E, latitude 25°, reported that, from $9^h\ 30^m$ to 10^h local time, the estimates were from 600 to 1000 meteors per minute. H. A. Newton, in 1886, collected all available observations of this shower, and wrote an important article[5] on the results. Among his conclusions were, that at one place, if all the sky were observed, the hourly rate would have been 75,000, the densest part of the stream was not over 100,000 miles in thickness; the principal shower did not last over 6 hours, in the densest part there was one meteor to every cube 20 miles on an edge; the radiant was an area several degrees across. The mean of 90 radiants gave $\alpha = 24°.54$; $\delta = +44°.74$.

The shower next appeared, but with greatly diminished brilliancy, on November 23, 1892 From the accounts gathered by H. A Newton into a short article,[6] the largest number reported seems to have been about 6 per minute. American observations only are referred to. On November 27 also, two travelers in Mexico reported very large numbers of meteors. It is very certain that this return was far inferior to those of 1872 and 1885

Finally, on November 24, 1899, in America and Europe, an even smaller shower appeared. The greatest rate reported was 90 per hour at Vienna, and 2 or 3 per minute at Princeton, N. J., for example. A considerable number was also seen on the previous night at Vienna. The radiant determined at Princeton was $\alpha = 23°$, $\delta = +42\frac{1}{4}°$, but was an area 2° to 3° in diameter. These details

[4] *Am Jour Sci*, (3), **31**, 81, 1886
[5] *Am. Jour. Sci*, (3), **31**, 409, 1886
[6] *Am Jour. Sci.*, (3), **45**, 61, 1892.

are taken from a résumé by W. F. Denning,[7] which he closes by remarking: "The earth appears to have passed through the extreme end of the shower at its recent appearance, and a brilliant return may be expected on November 17 or 18, 1905, when we are likely to encounter a region of the swarm much nearer the remains of the parent comet." The writer, in Virginia, purely by chance, happened to see something of this 1899 shower, 75 meteors in all being noted in an intermittent watch of 2 hours. The meteors were faint, and generally with short paths. The radiant area determined was very diffuse.

From 1899 to the present the Bielids seen have been so few that frequently no radiant could be determined, or, if one could, it rested upon a small number of meteors. Perturbations have shifted the orbit of the swarm, and doubtless have at the same time scattered it through a large volume of space. Any prediction that the shower will or will not return in future must depend very largely upon conjecture. Certainly we would overstep the bounds if we asserted that there are no chances whatever of seeing it again.

That the reader may understand more clearly why the changes mentioned have taken place, the following short table is copied from Newton's article giving the node and inclination of the comet's orbit, as calculated for the different returns, noting that the last two never were checked by observations.

YEAR	Ω	ι	YEAR	Ω	ι
1772	258 7°	17 1°	1846	246 5°	12 6°
1806	252 4	13 6	1852	246 3	12 6
1826	251 2	13 6	1859	246.1	12.4
1833	249 0	13 2	1866	246 0	12 0

The sun's longitudes (Equinox 1885.0) on the dates of displays are as follows:

1798	256 2°	1847	257 7°	1872	246 1°
1838	256 1	1867	248 4	1885	245 8

[7] *Monthly Not*, *R A. S*, **60**, 374, 1900.

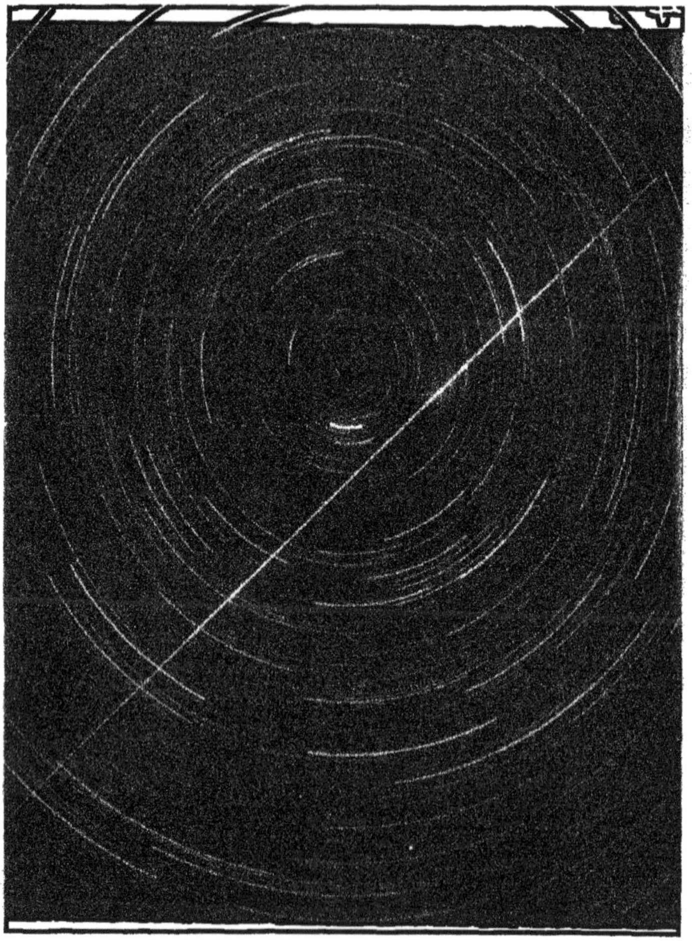

THE GREAT METEOR OF NOVEMBER 16, 1922

Photographed by W J S Lockyer, between 8^h 58^m and 11^h 12^m G M T, at the Norman Lockyer Observatory.

It will be seen that the first three nodes of the shower fit the 1772 orbit, while the last three fit the 1847 orbit. The comet approached Jupiter and suffered great perturbations in 1794, 1831 and 1841–2. Newton concluded:

> It is very difficult to conceive any way in which the meteors of 1872 and 1885 could be sent around in such a thin stream as we encountered, and one so near to the plane of the comet's orbit, unless they were a very compact group, and were also very near to the comet, as late as its near approach to Jupiter in 1841–1842. If they left the comet before that time the effect of Jupiter would not be the same on the meteoroids as on the comet, and if they at that time formed an extended group, such disturbances as the comet suffered would have scattered the group, and we should have had a much less brilliant star-shower in 1872 and 1885.

Newton also believes that the process of disintegration of the comet was quite rapid. It will be of interest to see how far the comet's orbit at the various returns actually missed intersecting that of the earth. The following distances, in terms of the astronomical unit, were computed by Hind[8] in 1872; + meaning that the comet's was outside, – that it was inside the earth's orbit at the point of nearest approach.

1772	–0 06545	1832	+0 00087	1852	–0 01130
1806	+0 01321	1839	–0 00009	1859	+0 00567
1826	+0 00892	1846	–0 01680	1866	+0 01295

As in 1846 and 1852 two nuclei were visible, the result is for the mean of the two possible distances.

Since comet disappeared after 1852, to push such computations to later returns than those given would be impracticable, or at least useless. The considerable changes in the elements of the comet's orbit keep us from being surprised that the history of the possible showers from it run back only to 1741, so far as known. These changes, the actual breaking up of the comet into two before our very eyes, and the splendid meteor showers which also were furnished by bodies moving in its orbit, have all combined to make it the most interesting case of all sighted to the astronomer. We can only hope that future perturbations will again switch the group

[8] *Monthly Not*, *R. A. S*, **33**, 98, 1872

across our path, so that more can be learned of the processes at work and how far they have progressed.

The elliptical orbit for the stream, calculated by S. J. Corrigan[9] follows:

	ANDROMEDES NOV 27	BIELA'S COMET
Pos of App radiant	α = 23 7° δ = + 44 3°	α = 24 0° δ = +43 2°
Pos of true radiant .	352 0 = + 9 3	349 9 = +7 7
Longitude of perihelion	108° 16′	109° 40′
Longitude of node	245 57	246 29
Inclination	13 08	12 33
Perihelion distance	0 8578	0 8606
Relative velocity	12 miles/sec	12 miles/sec.

In this, Denning's radiant for November 27, 1885, based upon the mean of 33 radiants, was used, and it was assumed that the eccentricity was the same as for the comet's orbit. The agreement is seen to be excellent.

From causes that will be developed later (see page 265) we have more reason to expect members of the Bielid groups, if large in size, to survive their passage through our atmosphere and reach the earth as meteorites, than for similar sized bodies of any of the other streams so far mentioned, except perhaps from the Pons-Winnecke stream. Quoting again from Farrington's table, we find 23 meteorites credited to November and 20 to December. Of these 11 fell in the last 9 days of November and 7 in the first 7 of December, in other words, quite an apparent excess for the dates when the shower might have been visible. We cannot go too far, however, since, owing to the shifting date of the shower we would have to consider also the year of each fall to make the comparison really fair. Also it must be remembered that the Geminids, another abundant annual shower, begin by December 1, and their radiant also is distant from the meteoric apex, and so would have a chance to furnish falls of meteorites, following the same line of reasoning as that already given.

One case for the Bielids is certain,[10] so far as the date is concerned, that of the Mazapil iron meteorite which fell in Mexico on November

[9] *Sidereal Messenger*, **5**, 144, 1886.
[10] *Am. Jour. Sci*, (3), **33**, 221, 1887.

27, 1885, while the brilliant shower, already described at some length, was in progress. It fell about 9:00 p.m., on a ranch 13 km. east of Mazapil. A most excellent account of the phenomenon was given by its owner, Sr. Mijares, as well as several other Mexicans. They all saw the wonderful shower in progress at the time. Unfortunately, Sr. Mijares did not see the meteorite coming down, but only after it fell, still hot enough to glow. It weighed 3864 grams, and made a hole 30 cm. deep. Professor Bonilla, of the Zacatecas Observatory, visited the place within five days, secured the accounts, and confirmed all details. The chances for a meteorite falling on any given night are so small that many high authorities speak most confidently of this iron mass as being a piece of Biela's Comet.

CHAPTER VIII

THE HALLEY'S COMET METEORS AND THE PONS-WINNECKE'S COMET METEORS

Of all periodic comets there is no doubt that Halley's has attracted most attention, not only from scientists but also from the public. Its well known history, which reaches back certainly to 11 B C. and possibly to 240 B C., and its appearance at or near certain events of great historical importance, for instance in 451 A.D., 1066, and 1456, have both been contributing causes. It is needless to recall to any reader the great amount of excitement and the innumerable articles caused by its last return in 1909–1910. Therefore for this comet a research on any possible meteor group connected therewith should have special significance.

Certainly as early as 1868 some guesses were made that the meteors which appear late in April and early in May might be connected with Halley's Comet. Rudolph Falb, in 1868, published the first paper[1] on the subject. That his conclusions were totally erroneous, so far as proved by his data, was at once shown by E. Weiss.[2] The next attempt seems to have been that by A. S. Herschel in 1876, contained in a list of 57 cases,[3] which he denominated *List of Theoretical Meteor Showers of Certain Comets, approximately corroborated by Observations* He gives the date and the nearest approach of each comet's orbit to that of the earth, and also the theoretical radiant. For Halley's Comet he gives May 4, 0.06 astronomical units, $\alpha = 337°$, $\delta = 0°$, and 'b' = 15° (this 'b' practically means difference between observed and computed place) The best accord he could then find was an agreement of 15°. It scarcely needs comment that such an agreement was a positive disagreement, and instead of proving would only disprove any connection. He followed this paper by another[4] in which he gives much more data, but still cannot find a closer agreement than 11°, a discordance out of all reasonable bounds. It can

[1] *Astr. Nach*, **72**, 361, 1868
[2] *Astr. Nach*, **73**, 41, 1868.
[3] *Monthly Not*, R A S, **36**, 210, 1876
[4] *Monthly Not*, R A S, **38**, 379, 1878

hardly be justly claimed, therefore, as has been done, that his work of that date made the connection even remotely probable.

In May, 1886, W. F. Denning published a paper *Meteor Showers of Halley's Comet*[5] In this he reviews Colonel Tupman's observations, previously used by Herschel as mentioned above. Tupman's radiant was. May 1–3, 1870, $\alpha = 325°$, $\delta = -2\frac{1}{2}°$. This shower was said to have been a brilliant one. H. Corder[6] on May 4, 1878, from three paths only, found a radiant at $\alpha = 334°$, $\delta = -1°$. However he added[7] 3 other tracks observed April–May and got as a result $\alpha = 334°$, $\delta = -5°$. His adopted position thus brought him to within 6° of the theoretical position of Herschel, but the fact that he combined several nights' observations of only 6 meteors vitiates his results. In any case it seems strange that the addition of only 3 more meteors, which really belong to the same stream, could change an adopted position, based on three others, by 4° in declination, without changing the right ascension Denning also by projecting 45 meteor paths, registered by different Italian observers about 1870, from April 29 to May 6, secured a position $\alpha = 335°$, $\delta = -9°$. Finally, Denning himself determined the radiant from 9 conforming meteors, saying that if three more were added the radiant may be diffuse to the extent of 5° or 7°. These twelve meteors were observed, one on May 2, 1880, the other eleven on April 30, May 1, 3, 4, 5 and 6, 1886. His result based upon 9 (or 12?) observations made on 7 different nights, is $\alpha = 337°$, $\delta = -2\frac{1}{2}°$. He concludes. "It agrees so closely with the radiant of Halley's Comet . . . . that the identity of the two orbits seems placed beyond doubt."

Thirteen years after this time, in 1899, Denning published his *General Catalogue of the Radiant Points of Meteor Showers, etc.* (see page 87), a work for which all future generations of astronomers will owe him a great debt of gratitude In referring to the group No. 258, called γ Aquarids, on page 223, in the "Remarks" column he says: "probably associated with Halley's Comet " Would it be unreasonable to ask why in 1886 the identity was placed "beyond doubt," while in 1899 the remark is "probably associated"? What had changed meantime? From page 283 we copy all the 8 radiants on which his final conclusion rested.

[5] *Monthly Not R A S.* **46**, 396, 1885.
[6] *Observatory*, **2**, 103, 1878–9
[7] *Monthly Not. R.A.S.* **40**, 135, 1879-80.

NUMBER	RADIANT	DURATION	OBSERVER	METEORS
1	329°, −2°	April. 29, 1871	Tupman	8
2	335 , −9	April 29–May 5, 1870	Italians	45
3	325 , −3	April 30, 1870	Tupman	15
4	337 , −2	April 30–May 6, 1886	Denning	11
5	334 , −5	April–May	Corder	6
6	339 , −5	May, 1892	Corder	6
7	338 , −2	May 1-4, 1895 and 1896	Corder + Bl	5
8	325 , −2	May 2–3, 1870	Tupman	13

It is submitted that the only thoroughly scientifically observed radiants in this list are Colonel Tupman's three. The others are at best all approximations because the meteors were observed on several different nights, and it has been proved by Olivier[8,9] and Dole[9] that the radiant is in daily motion. Colonel Tupman also observed from a position farther south, on shipboard and presumably under very good conditions. We may add to the facts just brought out that so far as known Denning never attempted to compute and certainly never published any orbit which could prove once for all whether his or any of the other seven positions quoted by him would give elements resembling those of Halley's Comet. For these reasons the writer felt justified in publishing in 1914[10] that, in his opinion, neither Herschel's nor Denning's work had proved the connection of the Aquarid meteors and Halley's Comet. Very recently an article[11] by A. S. Herschel was found which contains details of further observations on these meteors. In this he gives a radiant $\alpha = 335°$, $\delta = -1°$, based on 5 meteors seen between May 1 to 10 in the years 1894 and 1899, but 4 of the 5 seen in 1899. Again he gives a second radiant $\alpha = 339°$, $\delta = -2°$, from 3 meteors seen on May 1 to 7, 1900. Details of the appearances of the meteors are given by which he felt surer of their identification While this evidence is very slender, and the positions, based as they are upon several nights' observations, cannot be accurate, as the radiant is known to move daily, yet had the writer been aware of this article when he wrote his first statement on the matter, he would have modified his language used as to Her-

[8] *M1*, 9, 1911 and *Pub Astr Soc. Pac.*, **22**, 141, 1910
[9] *Observatory*, **44**, 242, 1921.
[10] *M2*, 469, 1914.
[11] *Bul. Soc. Astr. de France*, **15**, 81, 1901.

HALLEY'S COMET MAY 7, 1910, 5 00 A M.

Photograph by Geo F. Paddock, Lick Observatory Station, Santiago, Chile.

schel's credit in the discovery of the connection between these meteors and Halley's Comet.

The writer failed in various attempts to observe these meteors in years previous to 1910, due to bad weather, etc., but in that year, while at the Lick Observatory, good radiants were obtained on May 4 and 11. The parabolic elements correspondong to each radiant were at once calculated and tabulated beside those for Halley's Comet [12] As the radiants were determined separately from 6 meteors observed May 4, and 5 to 6 meteors on May 11, the meteors on each night all being observed within about an hour, and as the radiants proved to be very sharp, he felt justified in announcing that the connection between the η Aquarids and the comet was first definitely proved. The two orbits mentioned are given below with 6 others. Further papers on the subject appeared in 1911 and 1912 The whole results to that date (1914) were collected in *M* 2, from which the following is copied·

> In the table the elements based on eight radiants, secured from 1910 to 1913 inclusive, are given They are calculated in two ways First by assuming the meteors had parabolic velocity, second, that they have the same major-axis as Halley's Comet, and are moving in elliptical orbits In spite of the fact that the comet's orbit never comes nearer than 4,000,000 miles to the earth's orbit, and that consequently the orbits of meteors we meet cannot be absolutely identical with the comet's path, yet the agreement of the elements is so close for both cases that no doubt can remain that these meteors were originally intimately connected with the comet It is of great interest to see how far from the comet's orbit some of these meteors actually move This distance was computed to be as great as 11,000,000 miles for May 11, 1910 It should likewise be noted that the meteors were still coming in 1913 in nearly as great numbers as in 1910, when the comet was nearest the earth We can have no better example of the process of slow disintegration of a large comet into a meteor stream. . . . .

In 1921 R. M. Dole and the writer working on a similar plan, secured four more radiants,[13] which confirmed the eastward movement of the radiant first discovered in 1910, and announced in 1910 and more particularly in 1911 in *M* 1.

[12] *Pub Astr. Soc. Pac* , **22**, 141, 1910.
[13] *Observatory*, **44**, 242, 1921.

NUMBER OF COMET	PARABOLIC ORBITS					ELLIPTICAL ORBITS							OBSERVER	WEIGHT	NUMBER METEORS
	ι	q	π	Ω	π-Ω	ι	q	π	Ω	π-Ω	log e	log a			
						162 2°	0 587	169 0°	57 3°	111 7°	9 786	1 254			
182	157 7°	0 656	149 2°	41 8°	107 4°	157 6	0 646	147 5	41 8	105 8	9 984	1 254	C P O	1	5
183	161 0	0 696	155 3	43 3	112 0	161 3	0 712	159 0	43 3	113 7	9 982	1 254	C P O	3	10±
184	157 4	0 682	154 4	43 8	110 6	157 2	0 673	152 1	43 8	108 4	9 983	1 254	C P O	1	3
166	166 2	0 677	154 1	44 0	110 0	168 1	0 669	152 1	44 0	108 0	9 984	1 254	C P O	2	5
185	160 8	0 616	147 5	44 7	102 8	160 6	0 605	145 8	44 7	101 1	9 985	1 254	C P O	3	11
186	161 6	0 745	163 7	45 3	118 4	161 2	0 741	162 5	45 3	117 2	9 982	1 254	N B	1	17
167	163 1	0 608	147 6	45 9	101 7	163 0	0 598	145 4	45 9	99 6	9 985	1 254	G H	3	30±
168	166 7	0 630	155 1	50 9	104 3	166 6	0 621	153 2	50 9	102 3	9 985	1 254	C P O	2	5–6
Mean	161 8	0 664	153 4	45 0	108 4	162 0	0 658	152 0	45 0	108 1	9 984	1 254			
Weight mean	162 3	0 654	152 3	43 9	107 2	162 3	0 649	151 1	43 9	106 0	9 984	1 254			

Finally in this connection the writer in *M* 1, published in 1911, called attention to the great similarity between the orbits followed by the η Aquarids and the Orionids of October

A similar method of proof was given by C. Hoffmeister, at present the leading authority in Germany in practical meteoric astronomy, in a paper[14] dated January 28, 1912. He probably was then unaware of the writer's previous publications as he therein makes no mention of them. His proof was based upon work done in 1910 and 1911 by members of the Bureau Central Météorique, an international organization of Europe, which for some years before 1912 was most active. This paper was followed by a second,[15] dated August 21, 1913, in which he carefully calculates orbits based upon his own radiants as well as those of the writer, making the improvement in the latter of including the zenith attraction The greatest value of which came out, however, only 41′, so that no serious harm had been done by its omission in the writer's article, yet an error of judgment is freely admitted in not including it By grouping the 12 available positions into three mean groups, Hoffmeister solved for elliptical elements, assuming the same length for the major-axes of the meteor orbit and the comet orbit This gave him three different orbits, of which he reduced the first and last to the node-position of the second and combined. His final corrected elements were

$$\Omega = 45° 13' \qquad \pi = 144° 54'$$
$$\omega = 99\ 41 \qquad e = 0\ 96682$$
$$\iota = 162\ 38 \qquad q = 0\ 59480$$

[14] *Astr Nach*, **191**, 251, 1912

[15] *Astr Nach*, **196**, 309, 1913.

These elements agree on the whole fairly well with those derived by the writer in M 2, which were given on the previous page, and may be considered the best available for this meteor stream. Hoffmeister's treatment of the subject, in this paper, could indeed serve as an excellent model for any similar research

Were the node of the meteor stream stationary the showers mentioned on page 62 of A.D. 401, 839, 927 and 934 might possibly have come from these meteors But it is known that the node of Halley's Comet regresses, hence we would by analogy have to look in May rather than in April for such appearances. The only showers in Newton's list which by chance might be Aquarids seem to be May 19 4, A.D 842 and May 12 4, 1158 (the date reduced to 1850 is used) As the maximum may now be considered May 4 ± 2, reduced to 1850, we would have May 3 ± 2, 1922 This would give seven days regression in 316 years and 9 days in 764 years--a very poor inter-agreement. In general calling the regression 16° in 1080 years, we would have an average d Ω = $-0.9'$ per year. For Halley's Comet in 1835 d Ω = $-1.5'$, in 1910 d Ω = $-16''$ = $-0.3'$, the means of which would be d Ω = $-0.9'$. The comet itself appeared in 837 and 1145, so that the showers were seen respectively 5 and 13 years afterwards. While it is seriously doubted whether what has just been brought out has any significance, it is noted hoping that it may stimulate further research by others along this line.

More details about the articles and observations have been given concerning the accord of the η Aquarids and their connection with Halley's Comet than usually would have been necessary. This was, however, done purposely, as Denning published a note[16] in 1915 which takes an absolutely opposite view of the whole matter from that stated here. The facts in the case have, therefore, to the best of the writer's ability, been clearly set forth so that the reader may form his own conclusion in regard to the question at issue.

In 1916 meteors moving in orbits similar to that of Pons-Winnecke's Comet were met by the earth during the latter part of May and June, a strong maximum being observed in England on June 28. This maximum lasted only a few hours, as was proved by observers in America, who wholly failed to see it, though observing for meteors on that very night. The connection of these meteors and the comet

[16] *Observatory*, **38**, 366, 1915.

was independently announced by Denning[17] in England and by Olivier[18] in America. However, Denning has never published any orbits or other calculations, while the writer published the orbits with a full discussion of the case. A part of the article on the subject by the writer is here given, with a few additions.

The comet itself was discovered by Pons in 1819, and rediscovered by Winnecke in 1858 It appeared in 1869, 1875, 1886, 1892, 1898, 1909, 1915 and 1921. As the comet belongs to Jupiter's family and at times comes near the planet, very great perturbation ensue. To show these changes, which indeed will help us the better to understand the general behavior of meteor streams and comets, elements are copied for the orbits computed at four of the returns.

YEAR	log a	log e	log q	i	π	Ω	P
							y
1819	0 4997	9 8781	9 8885	10° 43′	274° 41′	113° 11′	5 62
1858	0 4965	9 8778	9 8859	10 48	275 39	113 12	5 56
1909	0 5135	9 8462	9 9879	18 17	271 37	99 21	5 88
1915	0 4510	9 8460	9 9875	18 18	271 43	99 23	5 87

Were all the orbits between 1858 and 1921 to be copied they would show progressive changes after 1858. In those given we are most interested in q and i, the first because it was mainly due to its changes from 0.77 in 1858 to 0.97 in 1909 that the shower of 1916 appeared, the second as an example of how the inclinations also are changed by perturbations, in a very few years.

On June 28.5, 1916, the comet's orbit came within about 0 03 astronomical units of the earth, while on May 27, when apparently the shower had already begun, according to the observations of J. Koep and P. Trudelle, members of the A. M. S., by whose work the writer first discovered the connection between the comet and meteors, the comet's orbit was 0.16 from the earth. This distance is indeed a great one, yet when we see the comet's orbit itself being so radically changed, it is not very strange that meteors which met the earth 10 months after the perihelion passage of the comet, which took place in September, 1915, might have suffered very different perturbations from those of the comet itself.

[17] *Observatory*, **40**, 95, 1917, Nature, **97**, 388 and 457, 1916

[18] *H. C. O. Circular* No. **614**, 1916; *Monthly Not*, *R A S.* **77**, 71, 1916.

Orbits were at once calculated for all the radiants available, both those derived in England and in America, and a comparison of their elements, with those of the comet, both on the parabolic hypothesis and when the semi-major axes were assumed equal to that of the comet (in 1909), showed such significant resemblances that no doubt could remain that these meteors were connected with that body. These orbits were published in the article mentioned. They are not reproduced here, however, because increased experience has led to the belief that the elements might be improved by another treatment and by having certain omitted corrections applied. It is, therefore, hoped later to make this the subject of a new research. In the meantime it is preferred to consider the orbits already published as first approximations only.

Nevertheless, the case was a most interesting one, as it showed that new showers were caused by perturbations throwing the orbit of a comet nearer to us. In this way new showers which have never been seen before may occur at any time, just as we have seen similar perturbations often cause such shifts that we no longer meet meteor groups which in former times furnished splendid displays. It was hoped that 1921 would see a rich return, but with the exception of observation of some very faint meteors made in Japan[19] nothing was seen of it. A few meteors only were observed elsewhere at about the time of maximum in 1921 and 1922, but the main group evidently had paid only a passing visit in 1916 and had then moved on. Again it is impossible to predict as to further returns, but for these meteors they seem less likely than for the Leonids.

[19] *Observatory*, **45**, 81, 1922. (Abstract.)

CHAPTER IX

RADIANTS

Doubtless the reader will be surprised to find a chapter entitled "Radiants" when the term was defined and explained in Chapter II and has been used on nearly every succeeding page. Yet here as in many another case, what appears the most simple is in fact very complicated. The writer has no hesitation in affirming that, in his opinion, the lack of an exact definition of the word, and of a clear understanding of what properly constitutes a radiant, has introduced more false ideas and complicated or made useless more meteoric work than any other single difficulty met in pursuing the subject.

In the chapter dealing with the Leonids it was shown that several American observers on November 12, 1833, discovered the fact of radiation and fixed the Leonid radiant with considerable precision. They noticed that for the hours of observation on that night the radiant appeared, at least approximately, a fixed point among the stars, as seen by each observer. The obvious and correct explanation of the phenomenon as an effect of perspective was at once given by Olmsted

But this was as is now known the simplest case because the meteors came from a radiant near the apex When it was fixed by the American observers it had a small zenith distance. The meteors came from a compact swarm, were seen within a fairly short interval of one night, and the stream being retrograde in motion was little liable to disruption and dispersion as compared to a direct stream

Coming back to the fundamental question, two elements are present, the area element and the time element Simply stated these are: how large an area of the sky could we possibly consider as belonging to the radiant area of a stream and over how long an interval of days, or greater units of time, can we consider meteors, which appear to radiate from the said area, to belong to the said radiant?

Before attempting to answer these questions directly let us enquire what causes influence the decision. The motion of a meteor in the earth's atmosphere is the resultant of several veocities which can be represented by vectors. These are, in order of magnitude, the velocity caused by the force due to the sun's attraction plus the original velocity of the body when it came within the sphere of the sun's attraction, velocity of the earth itself in its orbit, that due to the earth's attraction, and the small aberrational component due to the fact that the earth rotates on its axis daily We must add the unknown effect of the resistance of the atmosphere The first two velocities are known if we know the elements of the conic in which the meteor originally moved before it came into the earth's sphere of attraction. The velocity of the earth in its orbit is exactly known for every point and is as a mean about 18 miles/sec. The attraction due to the earth itself, called the zenith attraction and which will be fully treated in Chapter XV, sometimes attains considerable values, particularly for meteors overtaking the earth The diurnal aberration is never a very large quantity. As for the atmospheric resistance, while we know that it exists, for the meteors in our upper atmosphere we are as yet wholly unable to allow for it. Hence it will be omitted from the present discussion

From what has been said we see that *n* meteors, all moving parallel in space and with equal velocity, on entering the earth's atmosphere at any given instant will all undergo exactly the same effects, and hence would really radiate from a point. But let a second group of *n* meteors enter an hour later. Meantime the earth has revolved on its axis 15°, hence the radiant's zenith distance has changed Therefore the zenith attraction and diurnal aberration terms will be slightly different. Also in one hour the earth has changed the direction of the radius vector joining it to the sun by about $1° \div 24 = 2\frac{1}{2}'$. Therefore at any given place the direction from which the meteors came must have changed by an amount which depends upon the resultant in magnitude and direction of all the vectors which make up its apparent motion. In other words the radiant after an hour's time cannot be absolutely the same point in the sky, or will generally be a point generating a looped curve, as it moves eastward (see page 170).

It then becomes necessary to formulate rules so that we shall have practical means of determining radiants, whose positions shall be known to the limit of accuracy of the observations on which they depend. And it is now clear why both area and time interval come into the definition.

In justice to the observers before about 1870 it must be remembered that all these causes and their probable effects were only partially known. For those after the time of the appearance of Schiaparelli's book and who had access to it or similar scientific discussions on the subject there is little or no excuse for ignoring these facts. This is said advisedly because strongly critical language must be used in discussing so many lists of radiants, due to the manner in which the observations on which they rest were combined. The reader should therefore remember that only those men who published such lists during the last sixty years acted in the face of knowledge easily obtainable. All consideration should, however, be given to those earlier pioneer observers on whose labors depended future advances and whose very mistakes showed to their successors the better way for advancing.

From what was said in the discussion of the Perseids it appears that Locke in 1834 and Schaffer in 1837 both determined radiants for that stream. The first speaks of it as a "converging point," and makes no allusion that would cause us to think the radiant had any large area. The second speaks of the "center of radiation," and that he found it harder to designate this radiant point than that for the November meteors. As we saw, Herrick in 1838, while inclined to admit the radiant was north of that for the Leonids, confessed he was not sure of its position, much less anything else. Indeed he as well as most others at the time had difficulty in realizing that there might be many radiants in simultaneous activity, so a few meteors not conforming to the principal radiant for the night caused them great trouble. It was vain attempts at that period to force every meteor seen on a given night into the same radiant that hampered advances in many ways. It had a tendency particularly to make a radiant area abnormally large.

The first catalogue we need mention, containing 84 radiants, was published by Heis in 1867.[1] He divided the year into 22 periods, usually from the first to the middle, and middle to the last of each month. We can only conclude therefore that each radiant was the mean of observations obtained on from 1 to 15 nights. He does not inform us as to his radiant areas, whether they were large or small. As it is obvious that many nights' work was combined to

[1] *Astr. Nach.*, **69**, 157, 1867.

obtain the data for most of the radiants, these latter can only be considered as approximate and doubtless in unconfirmed cases as non-existent.

The second catalogue,[2] that of Schiaparelli in 1871, based upon the observations of Zezioli 1867–1869, inclusive, was made in a way that had others followed his example few difficulties would have arisen. Zezioli observed 7000 meteors in three years, yet Schiaparelli felt that he could only deduce 189 certain radiants from them. He states that only about one-fourth of the observed meteors belonged to these 189 positions He states also that usually only one night's observations were used to form a radiant, that in general at least five meteors must diverge from one and the same point for it to be considered a radiant, and that only meteors whose paths began near the radiant were given great weight. He calls attention to the fact that in several cases the same radiant appears on another night, generally in a slightly different position, and hence there are not 189 different streams represented in his list Further he calculated the parabolic elements corresponding to each radiant point. This necessitated the calculation of L, the longitude of the meteoric apex. It may be said here that the truly accurate way to compare radiants of different years is not by using the calendar date but by comparing the respective values of L. Schiaparelli thus gave us a model catalogue, compiled in the truly scientific spirit, where accuracy is aimed at, not the piling up of inferior data. To the present no list of radiants has appeared which contains any very great improvement over that just mentioned.

In 1872 R. P Greg published *A General Comparative Table of Radiant-Positions and Durations of Meteor Showers.*[3] This list contains 132 radiants. He used all material available, including Schiaparelli's catalogue But having his own ideas as to radiants, instead of copying in Schiaparelli's splendid 189 radiants, he grouped them into 79 "positions!" The radiants in this list of 132 are so vastly different in value that the work has to be carefully assorted. For instance where he was unable to combine two of Schiaparelli's positions, he merely copied down the originals unchanged. But while copying Heis's positions, which depended already on intervals up to 15 days as we have seen, in many cases he "confirmed"

[2] *Sternschnuppen*, 84–101

[3] *Monthly Not*, *R A S.*, **32**, 345, 1872.

the positions by other work which pushed the intervals from 15 to over 60 days! In his introduction he frankly states that in many cases the radiant areas are from 10° to 20° in diameter. He states that "the results are frequently remarkably coincident, when we consider how difficult it is to arrive, even in the case of the more frequently observed and better known Leonides and Perseides, at a precise radiant-point." The writer here begs to remark that no observer of any experience has the slightest difficulty in securing a sharp radiant for the Leonids, and even for the Perseids the radiant area is not unduly large nor is there uncertainty in determining its approximate center. In radiants which might have diameters up to 20°, and might depend on intervals of time greater than 60 days, absolutely no confidence can be placed. The list is merely useful as a résumé of former work, but its positions are generally quite inferior in accuracy to those of the originals.

The very next year, however, we find another English observer publishing a catalogue[4] deserving great praise This was Captain G. L. Tupman who, while cruising in the Mediterranean Sea, observed 3800 meteors in 1869–71 of which he mapped 2000. He derived 102 radiants, far the greater part derived each from observations on a single night. Without making a definite statement as to how many meteors were used to each radiant, as a minimum, he makes it clear that several meteors close to the radiant were necessary before the point was adopted. In an appendix he gives additional radiants for the Perseid and allied streams. While there are many outlying positions, we can easily pick out a large number which show the progressive shift of the main Perseid radiant, which was finally stated clearly by W. F. Denning in 1888 though probably noted ten years earlier by that observer. An accompanying map in Tupman's article makes this progression clear at a glance. His whole list is also of special interest as it contains a large percentage of southern radiants. Tupman had the sagacity to follow Schiaparelli's example rather than that of his fellow countryman Greg.

Among others, on the Continent, J. Schmidt in 1869 gave out a list[5] which was based upon intervals mostly half a month or a full

[4] *Monthly Not*, *R A S*, **33**, 298, 1873

[5] *Astr. Nach.*, **74**, 49, 1869.

a

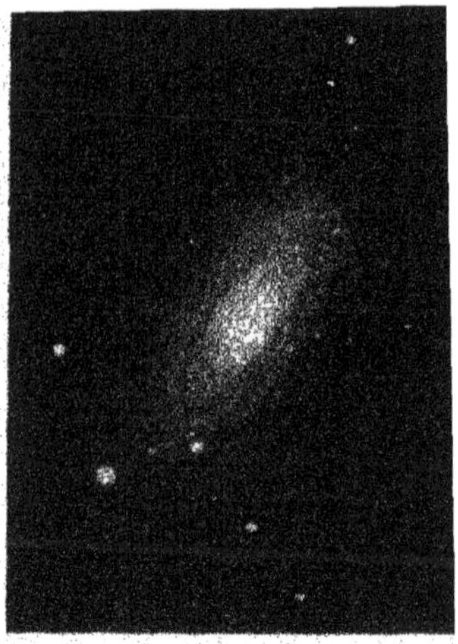

b

c

a, N G C 2841 Exposure 120^{m}
b, 2976 180
c, 2403 210

Taken by F G Pearse with the 60-inch reflector of the Mt Wilson Observatory.

month long. Heis's second catalogue appeared in 1877,[6] based on 14,000 meteors. It had a list of 36 radiants based upon stationary meteors, but the longer lists of radiants were based upon observations over varying intervals of time, and hence must be used with the same caution as similar ones already mentioned. The only other writer who made a distinct departure, which deserves comment, was Brédikhine, who carried out extensive observations in Russia, and wrote important articles[7] upon the theory of meteors. He avoided the error of combining observations of several nights, but had the curious idea that if he determined every intersection of two, three or more meteors' paths upon a map of a given region of the sky, say 40° by 40° in area, combining all these intersections, weighted according to the number of meteors belonging to each, then he would obtain the correct radiant for that part of the sky. At least this was his practice for the Perseid showers of 1893[8] and 1894. It indeed happens that this weighted mean usually fell near that radiant which was based upon the largest number of meteors. Nevertheless this procedure, in the writer's opinion, is quite untenable; and if those intersections depending on less than five paths were rejected and only the rest retained, then most of the actual radiants given by his observations would doubtless be preserved and the spurious rejected. It may be noted that the observations were made by men who presumably had not done very much meteor observing. This would partly explain the large number of meteors which were discordant.

The salient points, good and bad, having been pointed out in enough typical cases, we now come to the largest and most important work of the kind ever published. Its author is W. F. Denning and it appeared in 1899 in the *Memoirs of the Royal Astronomical Society*, 53, 203–293 under the title *General Catalogue of the Radiant Points of Meteoric Showers and of Fireballs and Shooting Stars observed at more than one Station*. As a work of reference for meteoric radiants there is no other comparable with it, and astronomers will be under the greatest obligation to its author for its compilation, which was a work of great magnitude

[6] *II Verof der konigl Sternwarte zu Munster*

[7] *Memoirs Académie Impériale des Sciences, St. Petersbourg*, 1890-1904.

[8] *Memoirs Académie Impériale des Sciences, St. Petersbourg*, 1, 33, 1894.

The catalogue contains 4367 radiants based upon about 120,000 observations, and, with the exception of some American, French, and Russian work, is practically complete up to 1899. Mainly for purposes of reference, as Denning says in the introduction, he divided the radiants into 278 groups which are placed in order of increasing right ascension, but all of the radiants of a group being held within limits of about 10° in both coordinates. There is no doubt that such a grouping is most advantageous for ready reference, and in this respect the catalogue could hardly be improved upon In regard to the actual radiants, as might be expected, they vastly differ in accuracy, as indeed he points out, because they represent all the work, good and bad, of which he found records

In reviewing the catalogue further it is necessary to speak of the very important part Denning's own work plays in its compilation. He began his meteor observing in 1866 and is still (1924) actively carrying it on. Even in 1899 he had to his personal credit over 14,000 observations, and in 1898 the Gold Medal of the Royal Astronomical Society was presented to him, chiefly on account of his meteor work. This long period of active and important service naturally brought him to the front rank of authorities on meteors, so much so indeed that from 1890 on his influence was predominant in Great Britain and very largely so on the Continent, at least in the observational part of the subject All this being so it is clear that his example was sure to be widely copied by others, who would feel very safe in following in his footsteps

It is therefore most regretable that with the practice of all the men, already mentioned, before him, he rather inclined to combine his observations in the manner of Greg and Heis instead of in that of Tupman and Schiaparelli. It is not intended to say that he did not very frequently determine radiants from observations made on one night, but certainly his general practice was not to do so. From this circumstance his radiants are necessarily of most unequal value, and indeed in a previous publication[9] he makes the statement that streams giving few meteors can hardly have their radiants determined within 5° or 7° of the correct position. However, it is obvious that owing to the great experience gained during his indefatigable work,[10] when he uses meteors observed on a single night his positions in general must be more accurate than this limit.

[9] *Monthly Not. R. A. S.*, **38**, 111, 1878.
[10] *Observatory*, **36**, III, 1878.

Finally, in regard to all work in the *General Catalogue*, no matter by whom carried out, the writer's opinion is that no radiant determined by meteors seen over the space of a month or more is worthy of the least consideration and that such radiant usually is non-existent, being the result of purely chance combinations. He must go even further and say that most of those, resting on the observations of over three consecutive nights, can at best be only approximate positions, while those based upon 3 to 30 days interval must either be for the most part the roughest approximations to the truth or, more usually, fictitious, due again to mere chance combinations. If this opinion is the correct one only a fraction of the positions in the *General Catalogue* are radiants in the strict sense of the word, while the rest fall under the criticism just given. Denning's influence shaped the policy of the British Astronomical Association's Meteor Section very largely, at least until 1920, and their older lists contain radiants frequently derived from very many nights' combined work. Meantime (1900–1920) meteor observing fell upon hard times in Europe, generally speaking, with the exception of the work (1908–1914) of the Bureau Central Météorique, and of a few individuals.

It was not until 1911 when the author published his first extensive work on meteors,[11] based upon 6200 observations of which 5000 were his own, that a serious attempt was made to return to the example of Schiaparelli and Tupman, and incidentally to launch a vigorous attack upon the system of combining observations, just criticised at such length. This was followed by the writer's second[12] and third[13] paper, in 1914 and 1920, based upon 2800 and 22,000 observations respectively, the work of the American Meteor Society and including 2000 or more observations of his own. In this way the writer published results based upon about 31,000 observations, mostly made under uniform instructions furnished by him. He also reduced all these observations. Over 1200 radiants were deduced therefrom.

More mature judgment forces the admission that many of these 1200 radiants, including a goodly number of personal ones, were

[11] *Transactions of the Am. Philosophical Soc.*, N.S. **22**, Part 1, 1911.

[12] *Publications of the Leander McCormick Obs.*, **2**, Part 4, 1914.

[13] *Publications of the Leander McCormick Obs.*, **2**, Part 7, 1920. For brevity these publications are referred to as M1, M2 and M3 respectively.

based upon too few meteors to be absolutely safe However, the great care in reduction, and the refusal to accept large radiant areas, and observations, in general, of more than one night mostly outweighed this error of judgement. Further, many of the weaker positions were found to be confirmed by those of other years, or by work elsewhere published whose accuracy was not to be doubted. Confidence therefore is felt that at least one-half of these 1200 radiants represent one or more appearances of real streams, and the writer will feel amply rewarded if the future proves this estimate not to be too high.

During the few years since the World War meteor observing has revived in Europe and good lists of radiants are again appearing That published by the British Astronomical Association in 1923 deserves special mention, for at last they have broken away from the practice of combining the work of many nights. Not only that but an advance of real significance has been made in including the correction for zenith attraction to the derived radiants, when this amounts to 0.6° or more. These, with other excellent features, make this list a model for all to follow.

Such free criticism of others in their ideas as to radiants having been expressed, it is but fair that the writer state his own views, along with the reasons for the same. This can readily be done by copying a report of the Meteor Committee of the American Astronomical Society, submitted in 1917, of which the writer was chairman. The report was prepared with the aid and approval of the following members of the Society: E. E. Barnard, W. L. Elkin, W. J. Humphreys, F. R. Moulton, H. A. Peck and W. H. Pickering. The report[14] was read before the Society and adopted in August, 1917. The sections dealing with radiants are as follows:

1a. A radiant shall be determined by not less than four meteors whose projected paths all intersect within a circle of 2° diameter, and which are all observed within a period of at most four hours on one night, by one observer.

1b. Or by three meteors on one night and at least two on the next night, seen during the same approximate hours of G.M.T., and all five intersecting as described above.

1c. Or by one stationary meteor.

[14] *Pop Astr.*, **26**, 18, 1918.

2a. A radiant shall be considered stationary for the period covered by observations when it fulfills (1a) for four consecutive nights and does not shift to an appreciable extent.

2b. Or when it fulfills (1b) for at least five nights or any longer period of time, provided in the latter case new positions are obtained at least every third night, for the whole period of its activity.

3a. A radiant shall be considered in motion or at rest when on examining successive maps of the same observer this radiant shall have moved to an appreciable extent, which from the accuracy of the observations can be considered unmistakable, or in the second place has kept the same position, the condition being the same.

3b Under no circumstances shall a meteor be used to determine a radiant-point whose projected path passes more than $3\frac{1}{2}°$ from the adopted point, and it is recommended that $2\frac{1}{2}°$ be generally adopted as the usual limit.

4. Three meteors which fulfill (1a) shall be considered enough to give a confirmatory radiant for one determined on the same date of a previous year—i.e., where L, the meteoric apex, differs by less than 2°.

5. No radiant shall be included in future catalogues which does not fall within the above conditions.

6 It is recommended that along with the date, which should be expressed in G.M.T. to tenths of a day, the longitude of the meteoric apex, L, should always be given. Further that a comparison of radiants in different years should be based upon L, rather than upon the date.

All the sections of the report bearing upon radiants have been copied though properly those dealing with stationary radiants should be included in the next chapter. Holding to these rather rigid definitions should in future: (1) Stop publication of fictitious radiants depending upon a few meteors scattered over long intervals, of which several old catalogues certainly contain many cases. (2) Make it possible to study the question of the motion or fixity of a given radiant point, with some degree of precision. (3) Obtain better radiants since the four hour limit will partly remove the effects of zenith attraction and diurnal aberration—the first of which is sometimes very great, particularly in the early evening hours. (4) Clear out from our future lists radiants depending on two or three meteors, unless there is some exceptional reason for retaining them. (5) Bring about the application of uniform rules to the work of all observers, thus tending to discard the poorer results and to retain only the better. (6) The placing of the whole question upon a scientific basis.

At present (1924) there is no means of knowing how many meteor observers in foreign countries have been willing to adopt these rules. However, they have been quite rigidly enforced as to the work in America, and the last report of the Meteor Section of the British Astronomical Association proves that they have been largely adopted by that Section. To this Association indeed meteoric astronomy owes very much, due to the great volume and general excellence of its work.

Two new methods of attack upon this problem must here be mentioned, the photographic and telescopic However only for the very best showers of the year can either method be expected to yield results. The photographic work done at Harvard and Yale, and already mentioned in Chapter II, gives information both as to the position of the centers of the radiant areas and their probable limits. The telescopic method was used successfully on August 10 and 12, 1921, by E. Öpik[15] of Dorpat, who observed the Perseid radiant with a five inch telescope, giving a field of 2° 32′ diameter.

He was able to map 15 telescopic meteors, and concluded that the area of the Perseid radiant, for August 12 when 14 of the meteors were observed, consisted of two oval areas whose (1855.0) positions were $\alpha = 42° 50'$, $\delta = +58° 15'$, area 3 3° × 1 3° by 5 meteors, and $\alpha = 40° 0'$, $\delta = +55° 35'$, area 5.7° × 2 2° by 9 meteors The meteors were from the 5 5 to 9.5 magnitude. His discussion is excellent, and the unexpected result proves the necessity for the employment of all optical means available to supplement the naked eye observations which are so liable to error. Further observations by observers with small telescopes and large fields of view are urgently needed and would bring out very important and interesting features about the radiants of such streams as the Perseids, Orionids and Geminids especially.

The numerical errors which might affect a radiant point by neglecting the various corrections and precautions mentioned will be fully taken up in Chapter XV

[15] *Astr Nach* , **217**, 41, 1922

CHAPTER X

Do Stationary Radiants Exist?

From the very divergent opinions held as to what constitutes a radiant one easily can predict that the answer to the above question largely will depend upon what definition is accepted. Aside from this fact, the question must be approached from two directions the observational and the theoretical. And, further, as the phenomenon of a stationary radiant is not to be expected except in special cases, observational evidence of great weight must be adduced before its sponsors can expect general acceptance of their results

While some of the earlier observers remarked that certain small areas of the sky furnished meteors over periods of varying length, and Schiaparelli himself called attention in a few definite cases to the recurrence of a radiant after a brief interval of inactivity in nearly the same place—for which, by the way, he gave an explanation—yet the claim that stationary radiation exists on what may be called a large scale is mostly due to W. F. Denning. This untiring observer published an article[1] in 1878 in which he states in a positive manner that his observations prove the existence of this phenomenon. From that time to the present, stationary radiation has been the most disputed point in meteoric astronomy, with Denning the leader of those who firmly believe in its reality It is thus inevitable that any discussion of the matter must lead to frequent reference to his extensive work

The second important article by him entitled *The Long Duration of Meteoric Radiant Points* appeared in 1884, and may be presumed to represent a later and maturer view than the paper of 1878 In this it is stated·

> The fact of stationary radiants exhibiting activity during several months is a phenomenon so unaccountable and so utterly opposed to the approved theories as to the orbits of shooting stars that it must receive most crucial examination before it can be accepted . Consecutive meteoric displays having no natural connection, though issuing from the same region,

---

[1] *Monthly Not*, *R A S*, **38**, 111, 1878

must necessarily exhibit small differences of position They will disagree to some 5°, 7° or more . But my observations absolutely prove that during several months the position of the radiant is unaltered . . . . This could not possibly be the effect were the displays a mere chance grouping of streams perfectly dissociated It has been said that even if a radiant point were found persistent for several months . the particles must pursue different orbits It would consist of a series of streams arranged in a peculiar manner, but not associated in a common orbit This may or may not be the correct interpretation of the fact.

He then states his belief that duplicate observations of meteors and fireballs are liable to probable errors of 5° to 10° in their deduced radiants, and stationary meteors fully 5°, while for a radiant determined by an isolated observer this drops to 2° or even 1°. Later are given four precautions to be followed of which the first is that the observations must be continued for as long a period as possible every clear night, and the last that observations of not more than two consecutive nights be combined.

Were observations indeed continued all night, or, in winter, for half the night, then the effects of zenith attraction become excessive for radiants whose zenith distances change considerably in the interval and which are near the anti-apex. The extreme correction may be – 17°. Nowhere in Denning's extensive lists of radiants, so far as noted, was this fact ever taken into consideration. As for the other point, the most casual glance will show that in very many of his personally derived radiants his own rule was violated, in some cases to the extent of an eighteen day combination!

He states in his article that the only objection to short intervals is due to the extreme feebleness of the great majority of showers, which give one meteor only every 3 or 4 hours. He adds: "I have seen streams so attenuated as to afford only about one meteor in 6 hours." The statement is made that the absence of meteors from a radiant for several hours is no proof of its total cessation and that it may revive again later, as the particles need not be uniformly distributed along the orbit. Also that stationary, long-continued radiants appear to be a general feature of attenuated showers, and that the most significant fact in connection with the subject is that certain sharply defined points exhibit a numerous retinue of showers, while the spaces immediately adjacent are comparatively barren of such displays. Along with the text are the numerical details of six cases which he considers to be especially well shown. These are as follows:

	RADIANT α	δ	APPARENT DURATION		SHOWER	POSITIONS AVERAGED
1	30 0°	+36 0°	July 16 to	November 14	β Tri	23
2	46 0	45 6	July 6	November 30	α-β Per.	31
3	61 0	47 7	July 25	November 27	μ Per.	21
4	61 8	36 8	August 2	December 31	ε Per	26
5	76 2	32 6	July 23	December 27	ι Aur.	21
6	80 2	22 9	August 24	January 15	ζ Tau	25

Denning's belief in stationary radiation is strongly reiterated in his *General Catalogue* and in numerous articles which have appeared up to the present It seems also to be true that a number of the other English observers believe in it and feel that their results confirm his opinions. All this being the case obviously it is vitally important for meteoric astronomy that such radiation be proved or disproved, at least in enough cases to justify inferences by analogy as to the others

Denning's first paper had hardly appeared before Tupman, of whose work such favorable mention has been made, gave briefly what appeared to him a perfectly satisfactory explanation[2] of the phenomenon, at least to the extent its existence had been indicated by observation up to that date As his reasons apply equally well today, and since in part have been independently given by others, they will be stated. He also gave for the first time the conditions under which a radiant can remain nearly stationary for a considerable interval. These are, that the orbit nearly coincide with the plane of the earth's orbit; the motion must be direct; the perihelion distance of the central portion must be a little less than unity, and the position of the radiant must be about 90° from the sun at the middle time. If the perihelion distance is 0.8 the earth would pass through the central part of the stream twice at intervals of about 3½ months. At the first passage the radiant would be about 125° before the sun, at the second about 30°, i.e., visible just after sunrise. In all other

[2] *Monthly Not., R.A.S.*, **38**, 115, 1878.

cases the reappearance of a radiant after an interval indicates a distinct shower. He then says:

The results obtained by Mr Denning . . are probably to be attributed to the method employed in obtaining the radiants . . . . The observations of many nights were combined to obtain the apparent radiation, and the duration was supposed to begin and end with the first and last day so employed It is easy to obtain a radiant in any part of the the sky . . by collecting night after night meteors whose tracks, carried backwards, pass near the point in question . . . Until the durations of fixed radiants be established in a more certain manner than those in the paper before us, we need not endeavor to explain them

By the time Denning's second and more complete paper appeared (1884) his reputation as a skillful observer was sufficiently great to force the attention of astronomers Important theoretical papers on the subject were published by von Niessl,[3] Tisserand,[4] H. H. Turner,[5] A. S. Herschel,[6] Brédikhine,[7] W. H. Pickering,[8] H. S. Plummer,[9] and M. Davidson.[10] All of these papers except that of Herschel are mathematical in character, some attempting to give an explanation of how stationary radiants could occur, others on theoretical grounds attacking the possibility of their occurrence, except under most special conditions. These astronomers, with the possible exception of Herschel and von Niessl, based their papers largely or wholly upon the data given out by Denning, and not upon personal observations and experience.

The question having greatly interested the writer as early as 1904 he turned serious attention to its observational side. From 1911 to the present he has, in a number of articles, discussed it from the standpoint of observation, besides making several critical studies of the published results of others. Hence the opinions expressed by

[3] *Sitz. d Kai. Acad. d Wissen , Wien,* **83**, II, 96, 1880

[4] *Comptes Rendus*, **109**, 341, 1889.

[5] *Monthly Not , R.A S ,* **59**, 140, 1899.

[6] *Monthly Not , R A S ,* **59**, 179, 1899

[7] *Bul de l'Acad. Imp. des Sciences, de St. Petersbourg,* **12**, 95 and **13**, 188, 1900 and 1901.

[8] *Astrophys Jour ,* **29**, 365, 1909

[9] *Monthly Not , R A S ,* **81**, 131, 1920–1.

[10] *Jour., B.A.A.,* **25**, 32, 1915, *Monthly Not., R.A.S.,* **81**, 414, 1921.

him at the end of this discussion will be based upon about 35000 observations, 8000 more or less being his own, which have been reduced by him in the interval 1898 to 1923.

Before giving the results of the numerous investigators mentioned let us draw a simple figure which will clearly show the meaning of the term true radiant versus apparent radiant, remembering that all discussion so far has been about apparent radiants only, as they alone can be fixed by observation The small effects of zenith attraction and diurnal aberration are omitted from this discussion as indeed they can be applied as corrections only after an apparent radiant has been approximately fixed from observations.

Let the motion of the earth in a unit of time be represented by the vector OB, that of the meteor by CB, and at the end of this unit of time let them meet at B. It is easily seen by elementary principles of physics that while the true direction of the meteor's motion is represented in both length and direction by CB, yet to an observer on the earth it seemed to come from C′, with a velocity equal to C′B. This is known as its relative velocity, being that actually observed. The true radiant is therefore in the direction C from B, the apparent radiant is in the direction C′ from B, and the angle CBC′ represents the arc in the sky between the two. Now it is readily seen that the greater CB is with respect to OB, the smaller the ∠CBC′ becomes. And in the limit for CB = ∞, the ∠CBC′ = 0. It is further seen that as ∠OBC increases, ∠CBC′ decreases, until when ∠OBC = 180°, ∠CBC′ = 0°, for a meteor coming from the apex there is no change in direction, only its apparent velocity will be the sum of the earth's plus its own.

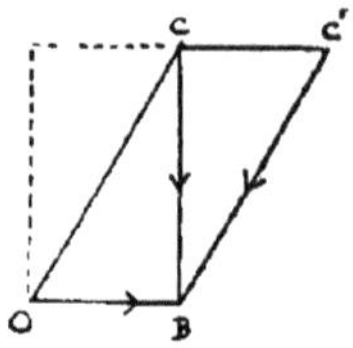

FIG. 3

From the principles just developed we are able to see: (1) That if a stream of meteors, moving in parallel paths with infinite velocity met the earth, the radiant would remain fixed, i.e., we should have a stationary radiant. (2) That the smaller the velocity of the meteors —CB in the figure—the greater the change in ∠CBC′ will be. (3) That the greater C′B is the smaller will be the change in ∠CBC′. (For the mathematical relations between apparent and true radiant see Chapter XV.)

All this was noted in the early years of the discussion by von Niessl in 1880,[11] and later Proctor in 1895[12] called in cosmical streams of great velocity to serve as an explanation of the phenomenon But Denning[13] at once replied that his observations proved that the angular velocities of the meteors from his stationary radiants, as well as those determined in some cases by duplicate observations of the same meteors coming from the said radiants, differed in no respect from those of the Leonids, Perseids and other groups, known to move with elliptical motion, whose true velocity, can in no case exceed 26 miles/sec. Hence Proctor's explanation instantly broke down, as we cannot accept Denning's data for part of his conclusions and reject the rest arbitrarily.

G. von Niessl then at Brunn, a man who developed theoretical meteoric astronomy in a masterly way, undertook in 1880 a really thorough research into the matter. The title of the paper is *Theoretische Untersuchungen uber die Verschiebungen der Radiationpunkte aufgeloster Meteorstrome.*[14] It has 48 full pages of contents, along with several most useful tables. Despite its great importance, the length of the mathematical discussion precludes any attempt here to follow out the actual equations, hence the arguments and conclusions only will be given as briefly as possible.

The paper was inspired by a desire to prove in what parts of the sky and under what conditions radiants can exist which will move so little daily that, so far as observations can show, they would appear fixed, within errors of observation. To do this the actual daily motion of radiants, distributed at certain chosen distances from the sun and the ecliptic, had to be computed and the results tabulated It was of course necessary first to derive the appropriate equations. In the first pages von Niessl discusses the question most fairly He states that since the catalogues at hand usually gave radiants, derived from observations of many nights combined, it was hopeless in most of the cases to prove anything which would be scientifically probable from such data. But that if the radiants from duplicate observations of fireballs and single meteors were compared and still gave nearly fixed radiants, a far better proof of its reality would be furnished. He states, however, that despite the generally poor data a

[11] Loc. cit.

[12] *Monthly Not*, *R A S*, **45**, 405 and 517, 1895.

[13] *Monthly Not*, *R.A S*, **46**, 444, 1895

[14] Loc. cit

few radiants seem to have been carefully determined and still to have remained fixed, and that his own observations bore this out Of these he gives no example, but contents himself with a table giving the radiants of 12 fireballs, observed from 1805 to 1872, for 11 of which he considered the radiant good. The mean of these, $\alpha = 59.0°$, $\delta = +20.0°$, he then compares with the mean of 10 radiants which gave $\alpha = 58.0°$, $\delta = +18.5°$.[15]

These 10 radiants were, all except the last three by Tupman, made in the usual manner of the day, the extreme time interval being that of Greg October 21 to November 30, the shortest that of Denning in 1872 October 29 to November 13.

Omitting the first (October 23, 1805), the other 11 fireballs were all seen from 1849 to 1872 between November 5 and 28, their

[15] It is of interest to search in Denning's *General Catalogue* for the details of the 10 radiants used by von Niessl In the table below is given first the table of von Niessl, secondly the data taken from the *General Catalogue* when they can be found

	α	δ	Gen Cat	α	δ	$\xrightarrow{\theta}$	Ref No
Greg 1875 Oct 21-Nov 30	60°	+19 5°	Oct 25–Nov 21	64°	+18°		20
Heis R4 Nov	55	16					
Denning 1872, Oct 29–Nov 13	62	20	Oct 29–Nov 13	62	20	13	22
Denning 1876, Oct 21–Nov 20	60	19	Oct 21-29	61	18	6	16
Denning 1877, Oct 2–Nov 13	60	20	*				
Corder 1876, Oct 21–Nov 30	60	18	Nov 21	62	21		74
Gruber ——, Nov 1–18	55	19	Nov 6	55	19		38
Tupman 1869 Nov 7	57	19 5	Nov 7	57	20	9	42
Tupman 1869 Nov 9	59	18	Nov 9	59	18	5	49
Tupman 1869 Nov 10–12	53	18	Nov 10-12	53	18	15	51
Mean	58 0	+18 5					

* In the *General Catalogue* this has been divided into three separate radiants, Ref No 13, 46 and 72 respectively, the mean of which is that given by von Niessl in first part of table The coordinates of Ref No 16 above and 46 just mentioned have been changed from those first published by him, for which see *Monthly Notices, R A S*, **50**, 456, 1890, and **38**, 110, 1877.

So far as these data go, which are those upon which von Niessl based his conclusions, it is submitted that a large part of them is so uncertain that it is not really useful for the discussion Several errors or inconsistences may further be seen in the *General Catalogue* In the latter it is but just to say the Denning has collected 84 positions in the group he calls No 53, the " Taurids The limits are $\alpha = 55°$ to $66°$, $\delta = 15°$ to $+23\frac{1}{2}$, an area 11° by $8\frac{1}{2}°$ Apparently he was unaware of von Niessl's paper on the subject, as the 12 fireball radiants by the latter are not included in the 84 radiants given observed heliocentric velocities being from 38 to 82 kilometers per second, mean 60 5 km/sec. He notes that of all the elements the velocity is most uncertain, and while if they all had a common origin they should have the same velocity yet these large deviations in velocities are not greater than errors of observation as to the exact angular length of path and duration of visibility. He further notes that the first 7 cases including November 13 would give as a mean $\alpha = 57.8° \pm 1.1°$, $\delta = +19.2° \pm 0.6°$, the rest from November 15 to 28 would give as a mean $\alpha = 60.5° \pm 0.9°$, $\delta = +21.0° \pm 0.5°$, but adds that he considers the radiants, even in these two means, too uncertain to give probability to the small motion indicated. He concludes this example by saying that it seems reasonable to believe that the near position of most of the radiants, both for the meteors and fireballs, can scarcely be merely by chance, and that if there is any motion of this radiant it is small enough to be wholly hidden beneath the errors of observation.

It may here be added that as the position is very near the ecliptic, if the angle $\lambda - \odot$ during the period of observations attains certain values then an almost stationary radiant could result for a month. The specific conditions will be considered later. Continuing his arguments von Niessl states that we may explain such an apparently fixed radiant in two ways (1) By assuming a number of different and unrelated streams to intersect the earth's orbit in such a manner that they will have the same apparent radiant. (2) By assuming a very wide stream, coming from space, whose different members have come from the same place or origin and whose orbits will therefore have the same aphelia. He hastens to add that for cometary, i.e., approximately parabolic velocity, only in very restricted cases could a stationary radiant result. However is it necessary to assume a cometary origin for all meteors? He calls up the fact that of 153 cases of well observed fireballs and large meteors, omitting Leonids, Lyrids and Perseids, of which he was able to find data, he found 23 cases or 15 per cent which did not equal the parabolic velocity, while 26 cases or 17 per cent were greater than 75 km/sec. The mean of the whole 153 was 61 5 km./sec., which is strongly hyperbolic. Further, the better observed meteors gave the higher velocities. Being also quite certain that he had proved fireballs came from the same radiants that furnished meteors he felt that the hypothesis of hyperbolic velocity for meteors, in many cases at least, was equally probable with that of parabolic velocity.

In developing the theory he assumes that the stream is made up of particles, vanishingly small in relation to their distances apart, and that the mutual attraction of one particle upon another is very small compared with that of the sun upon each. And that on its entry into the solar system any small part of the stream will have the same velocity and direction of motion. Further that the stream has a very great cross section, and hence that different parts must have different nodes, when meeting the earth.

Through a long process, which incidentally involved many of the equations found in Chapter XV on the derivations of orbits, he set up expressions which permitted him to determine, on any hypothesis of velocity whatever, the corresponding changes in the position λ, β of an apparent radiant and of a true radiant l, b. The interesting case was however when l, b was given and fixed, and it was desired to find the daily changes $\frac{d\lambda}{d\odot}$ and $\frac{d\beta}{d\odot}$. In this way it could be determined where and when, if ever, an apparent radiant appeared fixed, or, in general, how much it moved per day. As cases of special interest for $v = \sqrt{2}$ (parabolic velocity), when $\lambda_a - \odot = 144\frac{3}{4}°$ and $\lambda_b - \odot = 35\frac{1}{4}°$ a radiant in the ecliptic was stationary. As λ_b, β_b would in all cases set shortly after the sun, there is no great chance for meteors from such a radiant being observed. But λ_a, β_a would remain above the horizon most of the night. It was then calculated that if we take $\lambda_a = \lambda_2 =$ the middle of a month, $\lambda_1 =$ the beginning of the month, $\lambda_3 =$ the end of the month, and put $\lambda_2 = \odot + 144\frac{3}{4}°$.

$\lambda_1 = \odot + 143°$	$\beta_1 = 0°$
$\lambda_2 = \odot + 143\frac{3}{4}$	$\beta_2 = 0$
$\lambda_3 = \odot + 142\frac{1}{2}$	$\beta_3 = 0$

In other words for a whole month the shift in λ is only $2\frac{1}{4}°$ between extremes. As this is about the limit of accuracy when three observed radiants are compared, for such cases a stationary radiant may be said to exist. However for other positions of λ he found

For $\lambda = \odot + 0°$,	then $\frac{d\lambda}{d\odot} = +0\ 50°$
$\odot + 35\frac{1}{4}$	0.00
$\odot + 90$	−0.71
$\odot + 144\frac{3}{4}$	0.00
$\odot + 180$	+0 50
$\odot + 270$	+0.71

It is obvious that the average rate, even for a radiant with $\beta = 0°$, is about $\frac{1}{2}°$ per day. Further it is seen that for any value $\lambda = \odot + x$, then for $\lambda = 180° + \odot + x$, $\frac{d\lambda}{d\odot}$ has the same value but opposite sign. This investigation proved once for all, that even for radiants on the ecliptic, so long as v was not $> \sqrt{2}$, stationary radiants would not be very numerous.

Next the effect of increasing v to 2, 2.5 and 3 was investigated. For the stationary point $\frac{d\lambda}{d\odot} = 0$, values were found as follows:

For v = 2	$\lambda - \odot$ = 163.0°, or	$\lambda - \odot$ = 42.3°
2.5	167.5,	41.6
3	169.2,	40.7

These three values of v have maximum values of $\frac{d\lambda}{d\odot}$ equal to $\pm 0.88°$, $\pm 0.91°$, and $\pm 0.93°$ respectively but the important point is that as v increases the rate of increase of $\frac{d\lambda}{d\odot}$ near $\frac{d\lambda}{d\odot} = 0$ becomes smaller and therefore the chances for a stationary radiant increase.

As a test of the general correctness of the formulas, the Perseid radiant was assumed to be on August 10.5 at $\alpha = 44.0°$, $\delta = +57.0°$, or approximately $\lambda = 60.0°$, $\beta = +38.0°$ and $v = \sqrt{2}$. Then $\frac{d\lambda}{d\odot} = 1.12°$, $\frac{d\beta}{d\odot} = 1.12°$ and we obtain[16]

August	λ	β	α	δ
7.5	57.0°	+35.0°	42.0°	+53.0°
10.5	60.0	38.0	44.0	57.0
13.5	63.5	41.5	45.5	60.5

While space does not permit the reproduction of the full tables, an abridged one is here given containing just enough to show the general order of change of $\frac{ds}{d\odot}$ where $d\odot = 1°$, or about the daily motion of

[16] While the motion of the Perseid radiant has been fully confirmed by observations, yet the values so far obtained do not fit in very well with those calculated above.

the sun, $\frac{ds}{d\odot}$ representing the combined effects of $\frac{d\beta}{d\odot}$ and $\cos \beta \frac{d\lambda}{d\odot}$, i.e., the total daily angular motion.

λ—☉	V = √2						V = 2						V = 2 5		V = 3	
	90° 270°	120°	150°	180°	210°	240°	90° 270°	120°	150°	180°	210°	240°	90° 270°	180°	90° 270°	180°
β = 0°	∞	0 45°	0 09°	0 50°	0 65°	0 70°	0 50°	0 43°	0 13°	0 15°	0 33°	0 43°	0 40°	0 00°	0 33°	0 06°
10	∞	0 66	0 35	0 54	0 67	0 71	0 50	0 41	0 18	0 20	0 34	0 43	0 40	0 11	0 33	0 08
20	∞	1 24	0 66	0 63	0 70	0 75	0 50	0 46	0 26	0 25	0 36	0 43	0 40	0 17	0 33	0 14
30	∞	1 77	0 96	0 77	0 76	0 82	0 51	0 48	0 34	0 31	0 39	0 45	0 40	0 23	0 33	0 18
40	∞	2 22	1 22	0 91	0 85	0 93	0 51	0 51	0 41	0 36	0 43	0 16	0 41	0 28	0 33	0 23
50	∞	2 66	1 44	1 09	0 99	0 13	0 52	0 53	0 47	0 43	0 46	0 48	0 41	0 33	0 34	0 27
60	∞	3 17	1 76	1 32	1 23	1 51	0 52	0 54	0 51	0 49	0 49	0 50	0 41	0 37	0 34	0 30
70	∞	4 01	2 24	1 74	1 76	2 33	0 53	0 54	0 53	0 51	0 52	0 52	0 41	0 40	0 34	0 32
80	∞	6 77	3 77	3 03	3 30	5 04	0 53	0 54	0 53	0 53	0 53	0 53	0 41	0 41	0 34	0 34
90	∞	∞	∞	∞	∞	∞	0 53	0 53	0 53	0 53	0 53	0 53	0 42	0 42	0 34	0 34

A glance at the above table will show the essential results which are that the chances for an approximately stationary radiant. (1) Increases if the radiant is near the ecliptic (2) Increases the greater the heliocentric velocity of the meteors. (3) Increases the nearer the radiant lies to λ − ☉ = 180°. The converses of all these three conclusions are obviously equally true.

It is desirable to call particular attention to the fact that these results prove that no radiant for v< 3 can remain approximately fixed for as much as three months, and that as no one claims that v>3 for meteors from so-called stationary radiants, von Niessl's investigation proved absolutely that those radiants of several months' duration cannot be caused by physically connected streams of meteors, if indeed there is scientific observational proof that such radiants exist for any such periods

The next paper to be mentioned is Tisserand's which is interesting in that he took Denning's position of a certain supposed stationary radiant near β Trianguli, α = 30 0°, δ = +36 0°, and calculated the resultant orbits at five equidistant dates in its then-claimed period of activity. He found as follows·

Date	Perihelion distance	Inclination
July 20 .	0 891	140 4°
August 18	0 859	139 8
September 16 ..	0 293	111 3
October 15	0 305	39 8
November 13	0 704	15 6

It was seen that only for the first two dates could the orbits possibly refer to streams physically connected. Similar and much more extensive researches upon certain of the stationary radiants of Denning have since been carried out by Brédikhine and the writer.

The next contribution to the subject to be noted here was by H. H. Turner in a paper which appeared in 1899, and a second in 1900, the latter inspired by a lengthy criticism of the first one by Brédikhine which has meantime appeared and which will be mentioned on page 106. As the second paper contains a résumé of the conclusions of the first, as well as extensions, attention may be concentrated upon it. Turner built up his theory as follows. The earth is assumed to be at rest and its sphere of influence means the spherical volume of space around the earth throughout which a meteor is appreciably deviated by the earth's attraction. The following three propositions are then given as a further basis: (1) The magnitude of the velocity of the meteor on entering and leaving the sphere is the same. (2) The direction of motion is altered, being deflected in a plane through the earth's center. (3) The time of passing through the sphere of influence is shortened, because the velocity of the meteor is always increased and never diminished by the earth's attraction. Then by assuming a pair of passages through this sphere of influence, but during which the meteor passed on opposite sides of the earth, we have: (1) The velocity will be unchanged in amount. (2) The deflection will be = 0, being equal and opposite in the two cases. (3) The time of passage within sphere of influence is still further shortened

If these are true we should have a slow retrograde shifting of the nodes of certain meteors of a swarm, relative to others less affected, the relative velocity remaining the same. This would obviously tend to form a succession of nodes for members of the same stream, and Turner conceived these would thus furnish material for a stationary radiant. It was shown also that the conclusions were equally valid for a moving earth, and numerical results for certain interesting cases were obtained.

Newcomb pointed out this objection. Admitting all the above . . . . it will take a very long time and many passages before there will be sensible distributions of crossing points. Meantime the perturbations of other planets will so shift the stream that a member which passed near the earth on one return would be sure to pass further off at another, while new members would take its place rela-

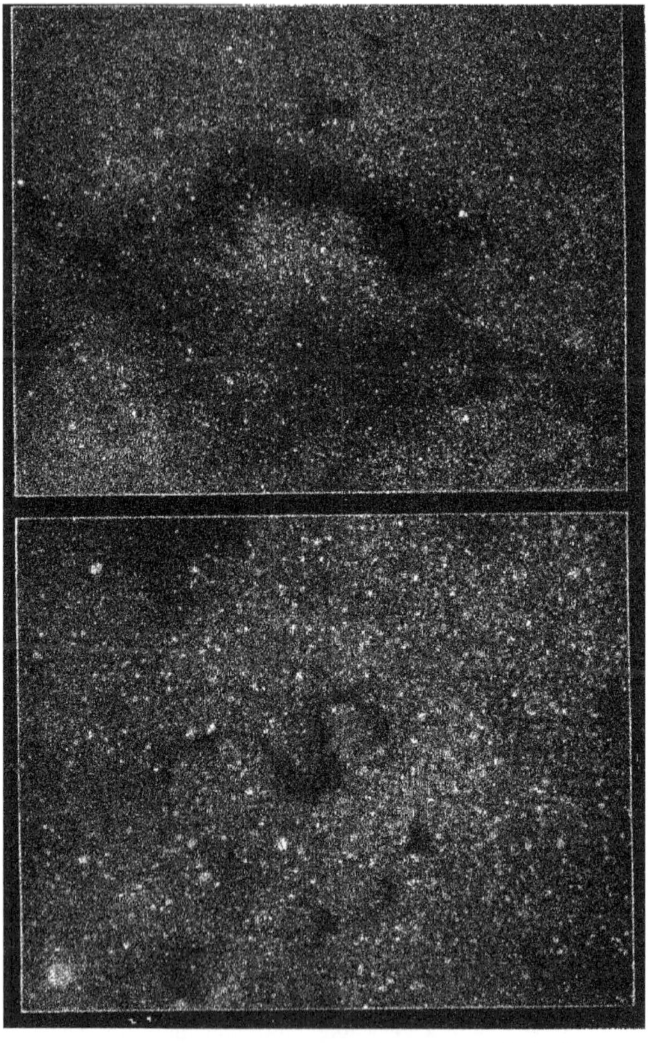

TWO DARK MARKINGS NEAR THETA OPHIUCHI BY E E. BARNARD AT YERKES OBSERVATORY

tive to the earth. Turner replied to this that if the parabolic velocity is assumed the deflection is certainly small. But are not the velocities probably smaller? And that there were many reasons why meteors should describe orbits much like the earth in which case the speed of distribution of the crossing points would be very rapid. He further suggests that the earth has an attractive effect upon meteors passing near it, hence concludes that numerous passages of the earth through a meteor stream would have a condensing effect upon its members. By means of a figure it is shown how this can be brought about.

It seems unnecessary to go into further details as two great difficulties remain unanswered. First we have no sound reason to believe that most meteors—least of all those from a stationary radiant—have velocities sensibly less than the parabolic. Second that while the earth does indeed exert such a condensing effect upon the outer members of a swarm it was long since proved by Schiaparelli (see Chapter XVII where the question will be discussed at length) that very many of the meteors which pass near the earth are thrown into hyperbolic orbits and hence never again meet the earth This last would of course continually deplete the stream. Unless therefore we are willing to admit velocities barely greater than 1, and much less than $\sqrt{2}$ as the usual thing, Turner's explanation can refer at best only to special cases, since it is wholly inadequate to account for the very large number of stationary radiants now declared to exist by Denning and others. Again since we have good reason to believe that as meteor systems are short-lived phenomena, cosmically speaking, as seen from the earth at least, it would be very easy to demonstrate that in any reasonable time the incommensurability of the periods of the earth and the meteors would permit too few encounters for the "pairs of encounters" to take place in any number.

Turner's first paper was immediately followed by one by A. S. Herschel in which he advanced an addition to Turner's theory because, according to it, as the relative velocity and radiant point would both remain fixed the apparent velocities of meteors from the same stationary radiant should also be the same all through the year. But unfortunately for the agreement of theory and observation such, according to Denning, was not the case. Herschel therefore assumed that the earth was originally surrounded by a ring of bodies, like Saturn's rings perhaps, and that in bygone ages streams of cosmical matter, moving with very great hyperbolic velocities, met the earth's ring from time to time. The result of the encounters would be to throw particles from the ring into orbits in which the relative directions would be almost the same and parallel to that of the cosmical current, and which might eventually return as meteors or even comets. Such meteors would, for each different cosmical current, give a stationary radiant. The assumptions on which this theory rests appear to be too improbable for serious discussion.

Turner added a third paper in 1902, fortified by 5 unexpectedly small velocities deduced by Elkin from the first Yale photographic results which had meantime appeared in print[17] These small observed velocities Elkin indeed attributed to the resistance of the earth's atmosphere. Turner, assuming a great height for the atmosphere, discussed the effects on a meteor's passage through the outer portion and deduced that the resultant resistance would lengthen the time the meteor remained in the sphere of influence This in turn would shorten the time necessary for the distribution of the crossing points of individual members, and would thus strengthen his theory. Unfortunately however for the validity of these deductions, later and more numerous, though unpublished, results at Yale frequently gave velocities of the predicted size and some even hyperbolic.

Following each of Turner's papers, Th. Brédikhine, a Russian astronomer of reputation who died in 1904, published one in criticism and reply. Here we are little concerned with the controversial side, for Turner's attempted explanation of stationary radiants has already been shown to be one which could scarcely hold as a general theory. But Brédikhine's excellent analysis of Denning's data, which in the meantime had appeared in full in his *General Catalogue*, explains many of Denning's results so well that it deserves extended mention. He begins with a careful statement of Denning's published views, which have already been given on page 94, and calls especial attention to certain claims that the latter made. It will be remembered that Denning states that most streams (and about 50 are said by him to be in activity on an average night) furnish only one meteor every 3 to 6 hours and that the radiants for many tenuous streams have a diameter of 5° to 7°. Denning however claimed that despite

[17] *Astroph. Jour*, **12**, 4, 1900

these facts he personally had experience enough generally to separate a meteor so that it might be assigned to the proper radiant, by means of its appearance such as color, sharpness, train, velocity, etc. He thus believed that he derived radiants with an error of not over 1° or 2°. Brédikhine here pauses to make the very pertinent remarks that apparent angular velocity depends upon the angle made by the meteor's actual motion with the line of sight, and that color can only be detected in the brighter meteors. He then quotes the case (from Denning) of the so-called $\alpha - \beta$ Perseids, where 18 radiants were given from July 6 to November 14, and where according to Denning's own observations the 15 cases have extreme ranges of 9° in right ascension and 7° in declination He then asks the unanswerable question why is it that, if these 11 radiants form a stationary radiant, and Denning is able to deduce radiants from his own observations with an error of not over 2°, there is an area 9° by 7° over which a "stationary radiant" can extend? Denning's own words that may such radiants are very sharply defined and not areas could also be quoted

The argument is continued by computing the parabolic elements for each separate position of the 18, of which we will only quote the inclinations in order to show the utter absurdity of there being any physical connection between all 18 cases 8°, 125°, 132°, 137°, 134°, 137°, 136°, 132°, 138°, 130°, 126°, 117°, 100°, 63°, 65°, 26°, 17°, and 10° The values of ι about 135° ± 5° in general belong to the radiants observed during the month of August. Several other cases given by Denning are treated in a similar manner and lead to similar results. He ends the discussion of Denning's work by the very obvious explanation that the various radiants observed were merely caused in general by successive streams which happened to appear to come from the same region of the sky.

In his second paper[18] he examines separately all the other 42 cases given by Denning for which the latter claimed stationary radiation. This was done rather to divide the radiants into two classes, the one formed by a simple and undisturbed current, the other by currents which had undergone strong planetary perturbations and been spread over an area, as is the case with the δ Aquarids which he had already discussed at length.[19] These he designated order I and II respectively

[18] *Bul de l'Acad Imp des Sciences de St Petersbourg*, **13**, 189, 1900

[19] *Bul de l'Acad Imp des Sciences de St Petersbourg*, **4**, 345, 1896

He then analyzed the 42 "stationary radiants" grouping the units of each into either single radiants of class I, or a group within a space of from 5 to 15 days into class II The rest of his article was taken up with replying to Turner's second paper, in which he designated the theory developed by the latter as "cette theorie née-morte," and meteors from stationary radiants as "ces météores élus" and "des météores magiques" But more seriously the following objection can scarcely be answered.

Dans leurs passages près de la Terre ils n'osent jamais s éloigner de la courbe même de l'orbite terrestre, en vain les perturbations causées non seulement par les planètes mais aussi par la Terre s'efforcent de les chasser de cette ligne, en vain cette orbite, moyennant le mouvement de son grande axe tâche de se défaire de leur extreme proximeté . Puis chacun de ces météores, dans ces rencontres consecutives, doit couper l'orbite terrestre tantôt en avant et tantôt en arrière du centre de la planète, à distances égales (en moyen) en évitent obstinément, dans ces conditions et durant un temps immense, de tomber sur la Terre, malgré l'invitation impérieuse de la théorie des probabilités

These papers of Brédikhine's will again be noted in the discussion of planetary perturbations. We may sum up the above contribution as follows If we grant that Denning's radiants actually exist as published by him, under no known theory of celestial mechanics could all the various single radiants grouped together by him to form a long enduring stationary radiant have any physical connection (with exceptions already noted by von Niessl).

W. H Pickering's paper in 1909 considered the question, using Denning's observations and frequently quoting Turner's articles, but in no case mentioning the very important work of either von Niessl or Brédikhine. He confines his discussions wholly to the case of radiants in the ecliptic, but later inferentially extends his results, without any proof whatever, in a general way to orbits of small inclination. He says

We may probably safely assume that the meteoric radiants are due, not to parabolic or hyperbolic orbits, but to ellipses whose major axes are of moderate length, and that the meteors themselves are in the majority of cases the visible remains, the ghosts, so to speak of defunct comets We may properly expect then that their orbits also will be inclined at small angles to the ecliptic

Large numbers of Denning's "stationary radiants" are far from the ecliptic, and a very large per cent of all meteor radiants, no matter by whom observed, are nowhere near the ecliptic. Also in tables of orbits, as those calculated by Schiaparelli, Konkoly, Kleiber, Bredikhine, Olivier and others, there are great numbers of retrograde orbits. The above assumption is therefore wholly untenable He comes to the conclusion that for (elliptical) orbits, when $\iota = 0°$, every radiant must remain within 16° of its mean position for eight months of the year, no matter how the orbit is constructed. Moreover no radiant in an unperturbed orbit can vary less than this amount, if it is observed continuously for eight months. And that "It is not likely, however, that any stream persists continuously for many months."

It is obvious therefore that even were W. H. Pickering's assumptions tenable—and we have just shown they are not—he can find no case for a stationary radiant continuing for many months. What the article really leads to is merely a very special case for elliptical orbits of the general case worked out long before by von Niessl

We next take up an interesting case, discussed by the latter author in his paper upon the great meteor of September 23, 1910.[20] The table contains data referring to fireballs, generally seen during the summer months, which radiate from the region of Scorpio.

NUMBER	DATE	APPARENT RADIANT		STARTING POINTS		VELOCITY km /sec	INCLINATION	ECCENTRICITY
		λ	β	l	b			
1	1908 May 19 9 7	253 0	+2 5	226 3	+2 8	35 5	°	
2	1869 20 11 3	242 1	+2 5	212 7	+2 6	64±		
3	1910 23 9 4	249 5	+3 3	220 8	+3 5	53 9	7 4	1 47
4	1886 June 2 11 0	251 3	+2 0	221 4	+2 0			
5	1899 2 8 8	250 7	−1 0	220 5	−1 0	Slow		
6	1883 3 9 8	251 4	+2 0	221 4	+2 0	47 3		
7	1883 3 10 6	249 9	+2 0	219 8	+2 0			
8	1878 7	250 9	+1 0	220 5	+1 0			
9	1873 17 8 8	250 5	+2 0	219 6	+1 7	28 9	3	
10	1908 28 11 2	242 0	+0 7	213 4	+1 1	28 2		
11	1908 July 1 9 2	251 0	+2 0	220 9	+2 0	29 0		
12	1901 8 11 6	254 5	−1 5	223 6	−1 1	19 3		
13	1879 13 8 4	247 9	+2 5	220 2	+1 7	40 2		
14	1900 17 8 8	250 6	+2 0	223 1	+1 3	Slow		
15	1872 22	250 0	+7 0	223 1	+4 6			
16	1846 23 9 5	247 9	−0 5	224 4	−0 3			
17	1903 Sept. 1	234 5	+4 5	224 3	+2 1			
18	1910 23 6 5	224 0	+5 0	219 4	+2 3	44 1	3 6	3 66

[20] *Sitz d Kai Akad d Wissen , Wien*, **121**, 1925, 1912

Von Niessl states that of the above 18 cases heliocentric velocities could be derived for 11 cases. The smallest was 47 km./sec., the largest was 80 km./sec., the mean velocity of the 11, unweighted, was 55.6 km./sec. For the three most accurately observed, No. 9, 13 and 15, he derived 69 km./sec. = 2.33 times the earth's velocity. The references have been looked up when possible and for completeness the hour has been added in the second column, and also the three columns for geocentric velocity, inclination and eccentricity. It is not clear in some cases whether the geocentric velocity has been corrected for the earth's attraction or not, or more correctly whether what is given is the true geocentric velocity or apparent velocity. These rarely differ by more than about a kilometer or two. Many of the original references were inaccessible

Due to the good accord of the values (l, b) in the fifth and sixth columns it appears indeed proved that most of these fireballs must have come from the same region of space, moving in parallel paths. The fact that they have been coming for 64 years or more also proves that this stream must have great width, otherwise the solar system would have passed through in a smaller time They are also able to furnish a nearly stationary radiant from the middle of May on for two months, but as their radiant lies nearly in the ecliptic it is one of the cases readily dealt with by the theory already developed.

In 1920 H. C. Plummer approached the problem in a somewhat different way. He first states that different kinds of explanations are necessary, according to how far the radiant is from the ecliptic, and that his investigations will be limited to the latter plane. He states that all evidence is suggestive that the orbits are large and therefore highly eccentric. His assumptions are that the meteors move in parabolic orbits with the same focus and same major axis, which he considers more nearly in accord with fact than the assumption that the paths themselves are parallel. He derives the following formula:

$$\tan\left(\phi + \tfrac{3}{4}\,\theta\right) = \frac{\sqrt{2} + 1}{\sqrt{2} + 1}\tan\tfrac{1}{4}\,\theta$$

where $\theta = \angle$ PSQ, in which P is position of earth, S is position of sun, SQ the common major axis. The angle ϕ is defined through the equations $R \sin\phi = -(V' - V)\sin\theta$, and $R\cos\phi = V' + (V' - V)\cos\theta$.

The longitude of the radiant can be expressed by o, θ, and ϕ. A table for every 10° of ϕ from $\phi = 0°$ to 180° was constructed containing various values of q, the perihelion distance, from 1.00 to 0.01, and ϕ calculated for both direct and retrograde orbits.

He at once concluded that for retrograde orbits no possible stationary radiant can be given, but that a displacement of about $\frac{2}{3}$° per day would be found for them, with remarkable uniformity. He then quotes the much discussed case of the Orionids (see p. 117) and concludes that "coaxial paths would give a displacement of the radiant amounting to 14° in three weeks." For meteors moving in direct coaxial orbits the maximum displacement would be 15.8° in three months, but "it (i e , 15.8°) is inconsistent as a whole with a stationary radiant." He further states that if at perihelion the width of the stream were 0.22, then a shift of only 4° would be given for six weeks, which might be within what he considers limits for a stationary radiant. A table of 6 selected positions of Denning, along with the latter's observed lengths of activity for each is given, but Plummer can in no case get anything more than a very partial agreement, the theory always giving far less an interval than Denning's. However he states that "The failure to account for the whole course of a stationary radiant does not affect the more limited result."

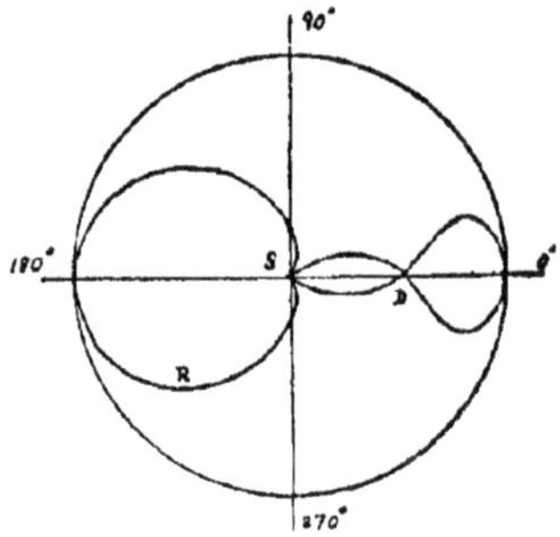

Fig. 4

In order therefore to see how a stationary radiant can be made to exist, he undertakes a different problem, still keeping $\beta = 0°$. This is solved by an arbitrary distribution of perihelia. Without following the equations, we merely give the figure, which contains the results. The perihelia lie upon the curious triple-looped curve, the heavy line part for direct motion orbits, the dotted for retrograde. His exact words are again quoted:

Stationary radiants in the ecliptic, which preserve their character for a comparatively short time at each appearance, seem to require no special explanation. . . . . The difficulty is associated with the long duration of certain radiants In order to explain an ecliptic radiant which lasts for six months at a time, the perihelia must be situated along the rather peculiar curve represented in figure 4. This is not a matter of theory beyond the simple assumption of parabolic velocity which is supported not only by general principles but by the descriptive notes of the observers, who have expressed a belief in the reality of stationary radiants in the wider sense . . . . the perihelia must have this peculiar arrangement. It is . . . necessary to explain the transition from direct to retrograde orbits

In conclusion he states that the conception of a stream of meteors must be widened. Also that in his article no dynamical theory is attempted, and no opinion expressed as to the real existence of persistent radiants in the widest sense. Yet for short durations they are certainly possible, as has been shown, and for most peculiar distribution of perihelia they are possible for long periods. Nevertheless in the closing lines Plummer admits that no physical explanation has been found, for it was as clear to him as anyone else that the chances that perihelia could or would be distributed along any such curve as that given by his figure are almost infinity to one. The paper, however, is a most welcome and important contribution to the subject and its method and conclusions appear sound.

Two instructive papers appeared in 1915 and 1921 by Rev. M. Davidson, a man to whose undoubted mathematical ability has been added the experience of an actual meteor observer—a combination seldom found! In the first he discusses what is meant by Denning's "radiation center of shooting stars," and says that possibly a cessation will ensue, and in a few months the shower will break out again from nearly the same place, and continue for another period. Certain criticisms of the writer's rules versus Denning's in the matter of radiants are given. But Davidson sets forth a new hypothesis in which he shows how a meteor stream could be flattened out, the larger particles pursuing their original orbits little changed, the smaller driven back into new orbits. This he conceives could be brought about by the sun's radiation effects, which would be greater for smaller particles (principle of ratio of mass to surface) than for larger. He deals with orbits in the ecliptic, direct motion, elliptical orbits with moderately long major axes and moderate eccentricities, and shows for $2a = 1.6$ and $e = 0.3$, that for a fortnight a shower may come from a small area 3° in radius; that after five months this shower may break out 8° from the first position; that the while the radiant gradually moves eastward, and that meteors will emanate for a part of the time from precisely the same position as before. Taking the radiant over a small area, we may say that it is stationary, but that the characteristics of the meteors may change. He feels that he is able to show that even for a whole year a stationary radiant could be explained by three original elliptical systems as described (the particles from each being forced to describe orbits with smaller semi-major axes and eccentricities instead of a continuous system of different orbits, as usually demanded by theory. Retrograde orbits are omitted from discussion, and the cases of radiants far from the ecliptic merely mentioned, no proof being actually attempted. The theory of how the particles, in an original system, of various sizes would ultimately be assorted into different orbits, the larger ones into orbits varying least from the original, follows the line of argument of Brédikhine's theory of comets' tails. It is certainly probable that such action takes place and its effects deserve further study.

In the second paper he reviews briefly those by Pickering, and Plummer, and again mentions one by Olivier, while he uses the results of Plummer in specific cases. For the parabolas assumed by the latter he substitutes ellipses with various values of a and e, and finds again that stationary radiants (always in the ecliptic) may exist for perhaps three months. Extending it further he is met with the difficulty that the inclinations begin to take values, in one case for instance, from 31° to 2°. A little later the motion becomes retrograde. Briefly his conclusion is the same as that of others—stationary radiants, with a liberal interpretation of the word, exist for not more than three months even in the ecliptic, so far as can be explained by any reasonable hypothesis or set of hypotheses.

Where then are we led? The most brilliant men who have turned their attention to the problem have offered us solutions only in particular, and, as is to be expected, the easiest cases. For general cases the explanations become so fantastic that no one seriously believes in their reality, their very authors advancing them merely as a step forward, which indeed they are, not as the final truth. So far as our knowledge now goes we must assume either large hyperbolic velocities, or that a succession of different radiants make up the average so-called stationary radiant, or that most of the published cases are due to errors of various kinds and do not accurately represent the observations on which they are supposed to be based.

CHAPTER XI

Suggested Explanations of Stationary Radiants

A detailed review of the more important papers, known to the writer, on the subject of stationary radiants having disclosed no general nor adequate explanation of the phenomenon, the writer desires to express his own views on the subject, as any light whatever in the present state of our knowledge is welcome. No new mathematical nor dynamical theory is here presented but, owing to his long personal experience in observing meteors and the great number of observations (about 35,000) that have passed through his hands, his point of view certainly is different from that of most men who have set up theories with little or no observational experience to guide them. From each predecessor will be taken that part of his theory which appears correct and, in the light of recent advances and personal observational experience, all will be fitted together into a reasonable whole.

As the observations of Denning, and, more recently, those of other members of the British Astronomical Association, are the ones most involved, it seems advisable to further analyze them as well as the other data in his *General Catalogue*, at the same time presenting those additional observational data in the hands of the writer which bear upon the questions at issue. Had some of the former writers on the subject taken the trouble to make such an analysis much labor would have been saved, but most of them not being meteor observers themselves accepted Denning's data and analysis without question As the writer has made a special study of three cases of 'stationary radiants', the α-β Perseids, the ϵ Arietids, and the Orionids, the discussion will be confined to these very typical cases According to Denning the Orionid radiant remains absolutely stationary for three weeks, the other two for the better part of the whole year.

The following evidence can be offered about the α-β Perseids. Out of 1224 radiants published by the writer[1] only 9 lie within 5° of the position $\alpha = 47.3°$, $\delta = +45.0°$, given by Denning[2] as the center of this area. Eight of these 9 were observed between August 8

[1] See *M*1, *M*2 and *M*3

[2] *General Catalogue*, 214.

and September 3 inclusive. The other was observed on October 31 by a member of the A. M. S., at that time wholly without experience. Therefore omitting this doubtful case, the extreme limit found is 27 days, with no radiants derived for the other 11 months. Personal work done between 1900 and 1909 was examined, 68 maps used in 6 different months being the material. On 24 maps no meteor whose projection came within 5° of the point given was found, on 18 one meteor, probably belonging to some other radiant, on 12 one meteor each, on 14 several meteors, but of these nearly all were used about August 11 and the meteors were merely Perseids, which happened to be in that sector. With fair confidence we may say that the work of the A. M. S. gives little or no support to any radiant at this point except for the month of August.

In 1919 Denning published an article containing 62 radiants of this 'shower', all but 8 from personal work, the mean being given as α = 48.2°, δ = +44 1°. Of these 54 however 19 were duplicate observations of single objects (see his own words quoted on page 94.) At least one of the 62 radiants falls in each month, but 15 fall wholly in August and 3 more partly in August as to time of observation. The extreme range is α = 44° to 52°, δ = +41° to +47° (omitting one extreme case based upon observations by others). The observations run from 1877 to 1918 inclusive, but only 9 of the radiants were observed in the first six months of the year. Of these 9 radiants 7 depended upon duplicate observations of one object. In the *General Catalogue*, 59 radiants were given including those of his own through the year 1898 and included in the above 62. Going back to these 62, it would be scientific procedure to discard 26 at once due to their depending upon observations over intervals from 6 to 29 days! This incidentally would remove the only 2 radiants observed before July, not including the 7 from duplicate observations. The only useful evidence he offers therefore consists of 19 duplicate observations of the same object of which he admits the probable errors may be from 5° to 10° in the radiants, and 17 radiants determined on from 1 to 5 nights' observations, of which 10 fell in August. As Denning's results covered over 40 years of observing and included quite 20,000 or more observations, the writer hardly feels that out of all this mass of work the evidence presented is very conclusive.

In all fairness it must be said that for instance out of 147 good radiants deduced from the work of the B. A. A. in 1922, No. 43, 68, and 119 were considered to belong to this group. They were based upon a duplicate observation on May 23, 7 meteors on August 30–31 and 5 meteors on October 17 Again Zezioli in 1867–1869 had observed radiants near this place on January 11, August 7, August 11 and September 18. Other observers could be quoted, if space permitted.

For this particular case, in view of all the evidence, these suggestions are offered in the presumed order of their importance and probability. (1) That during August this radiant is furnished by original members of the main Perseid stream, now moving in different orbits. (2) That from September 15 to 21 we have an entirely different stream intersecting the earth's orbit which will give the same approximate radiant. (3) That the same thing occurs about a month later. (4) That for the rest of the months no sound evidence exists for the supposition that a radiant could be observed each year in such a position. (5) That perhaps the fireballs seen in the spring months in part belong to one system. (6) That the rest of the radiants are fortuitous (as many must be especially if the observer expects to find a radiant in or near a given position) or are due merely to errors in observation or combinations of many nights' work.

Lastly are given the actual elements of certain streams whose radiants fall at or near this point, along with the observer's name Parabolic velocity was assumed in all cases.

DATE			α	δ	ι	π	q	Ω	OBSERVER	p	$\overrightarrow{s}$
1867–9	January	11	47°	+40°	9°	131°	0.97	291°	Zezioli		
1867	August	7	42	48	127	341	0.95	135	Zezioli	22	
1867		11	47	43	137	338	0.97	139	Zezioli		
1867	September	18	51	39	139	273	0.56	176	Zezioli	9	
1867	March	1	47	45	8	143	0.96	342	Denning by Kleiber	4	18
1877	July	20	47	45	125	7	0.78	118	Denning by Kleiber	5	1
1877	August	7	46	45	134	346	0.94	136	Denning by Kleiber	6	14
1877	October	5	47	45	100	242	0.32	193	Denning by Kleiber	14	11
1877	December	25	47	44	10	128	0.92	277	Denning by Kleiber	7	14
1901	August	8	42	49	126	339	0.97	136	Olivier	1	3
1901	August	9	44	40	143	299	0.99	137	Olivier	1	3
1913		10	45	40	143	301	0.99	138	Craig	1	12
1903		11	46	45	134	295	0.97	138	Olivier	1	3
1904		11	44	52	123	339	0.98	139	Olivier	1	5

A glance will show that these orbits, except the eight determined between August 7 to 11, have on the whole no resemblance and hence cannot all be physically connected

The Orionids, which have a maximum about October 19, furnish a very abundant shower and one for which it should not be difficult to determine the radiants as many meteors per hour are usually seen. The work of the writer and many other members of the A. M. S. has in general given evidence that the radiant moves in the direction of increasing right ascension. In 1922, during a most abundant return, R. M Dole, an observer of 25 years experience with many thousand meteor observations to his credit, strongly supported this evidence by 10 radiants secured each on a different night from October 17 to 29 inclusive A further support of unimpeachable character, since there can be no personal bias in a photograph, is from Harvard Observatory where Prof A. King, on October 20, 1922, secured on two plates 3 Orionid meteors, the radiant of which fitted in with the visual work of the American observers and showed a motion of about the expected amount from October 19, on which date all observers agree that the radiant is about $\alpha = 91°$, $\delta = +15°$. The details of the evidence may be found in the publications noted below.[3]

Several of the English observers headed by Denning have published observations showing two stationary radiants, one at $\alpha = 91°$, $\delta = +15°$, the other at $\alpha = 97°$, $\delta = +16°$, which Denning designates as Orionids and γ Geminids respectively, both of which are supposed to be in simultaneous activity. His point of view and that of others who support him may be found in his *General Catalogue* and the references below.[4] Mention may also be made of many articles in the publications of the B. A. A , particularly the reports of their Meteor Section.

As all the publications mentioned are recent, in English, and easily obtainable at any observatory, the arguments therein contained will not be set forth here. A few remarks, however, will be added on which the advocates of a moving radiant strongly support their conclusions.

1. The small observed motion fits with theory (see page 111) and has been shown by the records of several observers of experience in America.

[3] *M1*, *M2*, *M3*, also *Pop Astr* , **31**, 38, 1923, *Observatory*, **46**, 17 and **46**, 188, and *Monthly Not* , *R A S* , **74**, 37, 1913

[4] *Monthly Not* , *R A S* , **56**, 74, 1895; **73**, 667, 1913, *Observatory*, **41**, 60, 1918 and **46**, 46, 1923

2. The Harvard photographic position of October 20, 1922, confirms this motion. This was (for 1922 apparent place) $\alpha = 94° 26'$, $\delta = +15° 44'$.

3. Von Niessl gives (authority unknown) the following:[5]

Orionids:	$\alpha = 89.7° \pm 0.5°$	$\delta = +15.6° \pm 0.3°$	October 10–16
Orionids	91.5 ± 0.3	15.7 ± 0.3	October 16–22

This indicates some motion even though the observations were combined into two groups of six nights each.

4. The Orionids have retrograde motion ($\iota = 163°$) and to date no one has been able to set up a reasonable theory for such a stream to have a stationary radiant, unless we assume very excessive hyperbolic velocity for which there is not the least evidence.

5. Denning's first observations on the Orionids[6] were as follows:

> The radiant-point was found on October 18–19, by a projection of the paths on an 18-inch globe, to be R. A 92°, Dec +15°, on October 16 it was fixed at 90°, + 15°, so that it appears to have advanced with the time Mr H. Corder . . remarks that he "fancied the radiant shifted from 89°, + 18° on the 16th to 95°, + 17° on the 17th . . "

6. Denning's observations of the Orionids remained practically unchallenged until 1911. Up to that time little stress was laid by him upon the γ Geminids. But later when this latter was necessary to hold up the hypothesis by having *two* stationary radiants, we find the γ Geminids greatly stressed, and even said to be more abundant than the Orionids[7]

The writer therefore frankly states it as his opinion that so long as two "stationary radiants" are held to exist only 6° apart, and to be in simultaneous activity, but with greater activity during the latter part of the interval for the radiant further east,[7] the slightest predisposition to such a hypothesis on the part of any observer plus the unavoidable errors of observation would entirely mask the motion unless observations were continued throughout the whole period.

At the same time there is no doubt that minor radiants are, certainly during some years, in simultaneous activity very near the main radiant—a phenomenon all meteor observers who have studied

[5] *Smithsonian Misc. Collec* , **66**, 16, 25 (translation)
[6] *Observatory* **1**, 243, 1877
[7] *Observatory* **41**, 60, 1918

the Perseids should be familiar with. For instance a radiant has been noted a good many times by members of the A. M. S. which lies almost at α Orionis, and in any case there always will be a few non-conforming meteors which surely belong to the main stream, but which either before or after their entrance into our atmosphere are perturbed from their original paths. Briefly therefore the writer believes that the Orionid stream resembles the Perseids, but is passed through by the earth in about one-half the time, that its main radiant moves eastward daily, and that minor branches of the main stream (as is the case for similar streams) furnish small simultaneous showers, generally of an irregular character as regards annual appearance. All of these causes, along with unavoidable errors of observation and judgment, have given rise with many persons to the belief in two so-called "stationary radiants," neither of which the American observations confirm

As a final point the radiants of the Orionids lie very near the meteoric apex, and by the general theory of meteoric distribution we know that this, of all places in the sky, will probably contain most radiants per unit area. Hence in addition to the Orionids proper we should and do expect a number of inferior streams to have radiants in this neighborhood This indeed is found to be the case, though there is absolutely no reason why we should expect the same minor radiant to be annually detected, which proposition is again borne out by observations.

The last case, the ϵ Arietids, is especially interesting because it has been extensively studied by several other excellent English observers, who vouch for its long duration. Denning's evidence may be found in his *General Catalogue* and the *Observatory*, **42**, 246, 1919. Another most important article[8] by Miss A. Grace Cook and J. P M Prentice appeared two years ago. With regard to Denning's second article, 38 positions are given ranging from $\alpha = 38°$ to $46°$, $\delta = +16°$ to $+23°$. Duplicate observations are the basis for 15 of these radiants. The interval is from July 19 to January 25 inclusive. This article would be no further mentioned except that a quotation from it is necessary to prove that, as late as 1919 at least, Denning who is certainly the chief advocate of "stationary radiation" believed that such a list of radiants could have *one parent comet*

[8] *Monthly Not., R.A.S.*, **82**, 309, 1921, 22

Apparently there is no cometary orbit with suggestive resemblance to this meteor stream, but it may have been, and almost undoubtedly was, derived from a periodical comet which has long ceased to be visible as such, but has distributed its particles around the orbit and given rise to the annual meteoric displays that are witnessed.

This clear statement of his views was most fortunate as it completely silences those who, in attempting to strengthen his position, have frequently stated that he realized all the orbits corresponding to the radiants, included in one of his "stationary radiants," could not be enough alike to have had a common origin. In fact it comes very near contradicting at least one statement of his own,[9] if it does not actually do so. The series of observations by Miss Cook and Prentice present a much smaller range in both coordinates than those of any other observer, and, as they confirm Denning's in general, it seemed wise to undertake a mathematical discussion of the whole.[10]

As an approximate mean for the whole series reduced to the ecliptic system we have for the radiant $\lambda = 45°$, $\beta = +2°$, in other words almost on the ecliptic itself and therefore the most favorable possible case. In all 41 positions throughout the year were calculated, the parabolic velocity being assumed. Figure 4 shows the distribution of perihelia.[11] The inclination was $< 9°$ from October 11 to June 1, and $> 168°$ from June 21 to September 21, in other words was direct for about 8 months, retrograde for 4 months. The results may be briefly summarized by saying that when π, the longitude of the perihelion point, lies upon the two smaller loops of this complicated curve, we could have a more or less stationary radiant from October to May, during which also the motion is direct and the inclination nearly constant. But it is utterly impossible to suppose that the observations in July, August and September, when the motion is retrograde, belong either to the stream with direct motion or to any other single stream. In fact the very great differences between the longitudes of perihelia and the perihelion distances preclude any interpretation but that very many separate streams are involved. But unfortunately the observations of Miss Cook, Denning and Prentice all began in July and run to January. However the writer believes that for

[9] *Monthly Not.*, *R. A. S.*, **73**, 671, 1912–13.

[10] *Monthly Not.*, *R. A. S.*, **83**, 87, 1921–22.

[11] Figure 4 is not exactly the figure which accompanied this article but it is sufficiently similar for illustrative purposes

November, December and January his investigations give theoretical support to a more or less stationary radiant at this point. For the radiants which give streams with retrograde motions there can be no such explanation. Hence we are forced back to the types of explanation already given in the other cases, so far as these latter retrograde streams are concerned.

In the records of the A. M S. there are 10 radiants somewhere near this point. Of the 10 we find 4 from August 9 to 12 inclusive, which obviously prove that there is a stream with this radiant active at that date. The last 4 from October 22 to November 6 probably belong to the approximately stationary radiant, as during this interval our table shows slight changes only in the elements, with the inclination $\iota = 2°$ and about constant. The two remaining positions are No. 84, $\alpha = 50°$, $\delta = +23°$, which is too far off to really be a member, and No. 646, $\alpha = 40°$, $\delta = +26°$, observed August 24, 1917, and based on 5 meteors which fits in well with similar English observations. In conclusion the investigation of this stream proves that there is an active radiant in this position about August 9 to 12, and that the radiant is again active from October 22 to November 6; also that theoretically a stationary radiant may be expected for about three months beginning at the latter period. Any streams from this approximate position in August must be entirely different streams as they are obliged to have retrograde motion

On the basis of everything that has been pointed out, the writer's present opinions upon the whole subject are as follows: There is no doubt that for radiants near the ecliptic approximately stationary radiants may exist for considerable periods of time. But so far no satisfactory theory has ever been advanced for radiants with high latitudes and that are supposed to be stationary. How then are we to account for experienced observers vouching for their presence? Several reasons can be given some of which apply to certain individual cases, not necessarily all to each. First in importance stands the combination of observations made on many nights which can be proved mathematically to be unsound practice. Second no corrections for zenith attraction have been applied in 99 per cent of the published work. Third assuming radiants to have unreasonably large areas. For instance if a diameter of 6° is assumed, the projected paths of at least one meteor in every 60 has to pass through this area, whether it belongs to the said radiant or not. Fourth having fixed ideas that a radiant has been, hence ought to be, found in a given place. In other words it is a well-known psychological fact that persons often see what they are led to expect to see. In meteor observing, where the whole appearance of the object usually is over in a fraction of a second and the impression left not of the highest order of accuracy, such an effect must constantly be guarded against by even the most honorable and high-minded observers Fifth, even when none of these causes are present the inherent errors of observation, poor projection of many maps, poor observing conditions, physical fatigue and many other such reasons will cause numbers of spurious radiants to be found, many of which must coincide with the very great number of Denning's "stationary radiants"! The average area of the latter is at least 8° by 8°. It would therefore require only about 325 such areas to completely fill the northern hemisphere of the sky, where most have been reported to exist.

Looking at the matter from another standpoint, there is not the least reason to try to force every meteor, or perhaps even the majority seen on an average night, into radiants Not all meteors are parts of existing streams. There must be vast numbers following entirely unique paths in space. Hence when an observer derives too many radiants per thousand of meteors his work is open to immediate suspicion. For instance Schiaparelli deduced only 189 radiants from Zezioli's 7000 observations, and other good examples have been given. Again for streams like the Bielids which meet us from near the antiapex a few returns suffice to scatter their members so the radiant covers a very large area of the sky. What are the ultimate radiants for those most perturbed? They certainly will meet us in such different directions that eventually we could not even suspect they once were Bielids. Everything leads us to think that meteor streams, so called, especially those with direct motion and small inclination, are very temporary aggregations. Hence near the ecliptic, which is almost in the fundamental plane of the solar system, there must be large numbers of minor streams, long since dissociated from larger streams, and even vaster numbers of isolated units which are dissolved parts of the minor, so that the earth must meet in time streams of every degree of composition, and isolated bodies to an inconceivable number. Again for small groups in elliptical orbits there may be years between returns, or perturbations may meantime switch them out of their old orbit so that the earth meets them no more.

As most of this reasoning scarcely can be denied, it is clear that we should expect to meet innumerable isolated meteors whose "radiants" could be determined only by a chance duplicate observation, many small groups which would give a radiant one year but not annually, and many that never will return. This last will be vastly increased if we admit the existence of stellar swarms. With all this in view the scientific procedure is to cut down radiants to a minimum, admitting only those whose existence is proved by observers in different countries and different years, and those determined beyond question by one observer of experience based upon a fairly large number of meteors. But it must be remembered that actually, if we count small groups, radiants in very great numbers do exist, scattered all over the visible heavens, and that by uncritical combinations of existing material one can prove that almost anything in the way of motion of radiants or its opposite has been detected. A view somewhat similar to this last, has been clearly advanced by others, most recently Hoffmeister.

As a final conclusion the prediction is hazarded that with time so many excellently determined radiants will be observed and catalogued that one at least for some data or other will be found in nearly every square degree of the heavens And this after our more exact successors have discarded most of the relatively poorly determined material now at hand. Meantime our purpose will have been fulfilled if some of the true reasons have been given why stationary radiants, in the sense now accepted, are supposed by many to exist on a large scale, and also how, to a considerable extent, they may be explained as fortuitous and illusory. It is here desired to pay a further tribute to the analytical work of von Niessl and Brédikhine on this subject, for it is believed that had their work been readily accessible to every meteor observer we would be hearing very little now about stationary radiants as a widely observed phenomenon.

CHAPTER XII

The Apparent Paths of Meteors Within the Atmosphere

Were the earth without its atmosphere, the paths then followed by meteors obviously would be slightly different from those now described. The causes of this difference are numerous. First in importance is the resistance of the air which becomes immensely great to a body moving with a planetary velocity. Next, the deviating effect of the atmosphere upon any but an absolutely symmetrical body, a sphere for instance. Then the deviation due to the rotation of the atmosphere itself every 24 hours around the earth's axis, and, lastly, the effect of the density of different strata, and of the high winds.

The study of the resistance of the atmosphere to projectiles has been highly developed in the specialized branch of mechanics known as ballistics, particularly since the World War. Yet, so far, a velocity of from 1 to 2 km per second represents the limit of human achievements in artillery projectiles. We have every reason to believe that meteors could scarcely move this slowly (except when almost down to the earth's surface) and they may attain velocities fifty to a hundred times greater. Again we have little real information about the densities and gases in our atmosphere higher than 35 km. Sounding balloons have sometimes penetrated to this altitude. Observations on high mountains and by aeroplane have given information up to 11 km. Beyond 35 km, however, instruments have not been carried and our formulae are therefore exterpolated for the heights at which meteors appear and generally disappear.

The following description of the atmosphere above middle latitudes of our earth is copied from *Physics of the Air*, by W. J. Humphreys. We do not reproduce the figure given by him, to which this description applies, but only copy the table from whose data the figure is drawn. This table is of course exterpolated entirely for values above the upper limit to which it has been possible to send sounding balloons, i e., 35 km. Nevertheless it is based upon the best available interpretation of known physical laws and its results are generally believed approximately to represent the actual conditions.

Percentage distribution of gases in the atmosphere

	GASES							
Height	Argon	Nitrogen	Water vapor	Oxygen	Carbon dioxide	Hydrogen	Helium	Total pressure
km								*mm*
140		0 10				99 63	0 36	0 0040
130		0 04				99 55	0 41	0 0046
120		0 19				99 35	0 46	0 0052
110		0 68	0 02	0 02		98 97	0 51	0 0059
100		2 97	0 05	0 11		96 31	0 56	0 0067
90		9 86	0 10	0 49		88 97	0 58	0 0081
80		32 39	0 17	1 86		65 11	0 47	0 0123
70	0 03	52 04	0 20	4 74		32 73	0 26	0 0274
60	0 03	81 33	0 15	7 70		10 69	0 10	0 0935
50	0 12	86 82	0 10	10 17		2 76	0 03	0 403
40	0 22	86 43	0 06	12 61		0 67	0 01	1 84
30	0 35	84 27	0 03	15 18	0 01	0 16		8 63
20	0 59	81 24	0 02	18 10	0 01	0 04		40 99
15	0 77	79 52	0 01	19 66	0 02	0 02		89 66
11	0 94	78 02	0 01	20 99	0 03	0 01		168 0
5	0 94	77 89	0 18	20 95	0 03	0 01		405 0
0	0 93	77 08	1 20	20 75	0 03	0 01		760 0

. . . . The assumptions upon which it (i e., the table) is based are in close agreement with the average conditions of middle latitudes, and are as follows.

1. That at the surface of the earth the principal gases of the atmosphere and their respective volume percentages in dry air are:

Nitrogen.	78 03	Neon	0 0015
Oxygen	20 99	Helium	0 00015
Argon.	0 94	Carbon dioxide.	0 03
Hydrogen	0 01		

2 That water vapor at the surface of the earth amounts to 1.2 per cent of the total quantity of gas present.

3. That the water vapor rapidly decreases, under the influence of lower temperatures, with increase of elevation, to a negligibly small amount at or below the level of 10 kilometers

4 That the temperature decreases at the average rate of 6°C. per kilometer from 11°C at sea level to −55°C at an elevation of 11 km.

5 That beyond 11 km above sea level the temperature remains constant at −55°C.

6. That up to the level of 11 km., the relative percentages of the several gases, excepting water vapor, remain constant—a result, of course, of vertical convection.

7 That above 11 km., where the temperature changes but little with elevation, and where vertical convection, therefore, is practically absent, the several gases are distributed according to their respective molecular weights.

He adds that on account of their small quantities near the surface neon, krypton, zenon, ozone, etc., have been omitted from the table. Also he warns especially that, after all, the values in the table above 30 km. are extrapolated values only, and hence must become less certain with increase in elevation. Also that all hydrogen values are in doubt as no two observers get the same amount, even at the surface. Hydrogen appears to be shown in the spectrum of one meteor out of the three available. Yet we do not know whether the spectrum was from the hydrogen in the air or from that in the meteor itself. Hydrogen is not shown in the auroral spectrum. Some observers even claim that there is no free hydrogen at all in our atmosphere.

It will be of interest to state here that the latest work of C. Stormer[1] and L. Vegard[2] seems to prove that the spectrum of the aurora has the nitrogen lines, as well as some unknown lines, the former being observed to the altitude of about 500 km. Vegard assumes a dust atmosphere above the gaseous, the latter ending at about 80 to 100 km. above the surface, the former comparatively dense at 100 to 120 km., from which level its density slowly decreases as we go upward. His value for the bright green line is 5578. High authorities in meteorology, however, consider Vegard's assumptions as entirely incorrect. The recent, and almost revolutionary, conclusions of Lindemann and Dobson on the constitution of the upper atmosphere will be spoken of a few pages further on.

Another recent study of the green auroral line has been made by H. D. Babcock, with a Fabry and Perrot interferometer at Mount Wilson, Calif.[3] He finds the wave length to be 5577.350 ± 0.005 and states that the line may, if certain assumptions he made are correct, possibly be due to helium.

As the earth's atmosphere rotates at the equator with a velocity of 0.465 km./sec., and at the height of 100 km. with a velocity of 0.471 km./sec., or roughly $\frac{1}{2}$ km./sec., again taking the velocities of meteors at 15 and 75 km./sec., as extremes, we would have deviations from straight lines of about $\frac{1}{30}$ and $\frac{1}{150}$ respectively, i.e., about

[1] *Comptes Rendus*, **176**, 109, 1923.
[2] *Philosophic Magazine*, **46**, 193 and 577, 1923.
[3] *Contribution . . . Mount Wilson Obs.*, **No. 259**, 1923.

2° and 0.4°. Except in very long paths even the first of these values would be wholly unnoticed by the eye; the second could not be detected. These values decrease as we approach either pole. For the latitude ϕ we would have the value 2° cos ϕ. At the pole, $\phi = 90°$ and the effect becomes zero. No further attention need be given to this effect.

The other lesser causes are easily disposed of. As to the wind, let us suppose that it is blowing 120 km./hour, which would be vastly greater than usual. This would be only $\frac{1}{30}$ km./sec. Even very slow meteors coming from the anti-apex generally will move about 15 km./sec Therefore the ratio would be at most one part in 450 Expressing this in arc it would be less than 6′. Hence it is easily seen that we could never hope to detect any deviation due to the wind, as the above values were calculated for hurricane velocity, not the average winds.

It would lead into great digressions to take up the mathematical side of the behavior of irregular projectiles, though we must indeed assume that many meteors, if not most, are irregular in shape. A few simple illustrations will suffice to prove the actuality of curved or irregular paths of such projectiles in a medium like the air. The Australian boomerang is familiar to all. This projectile is shaped with one side almost flat, the other decidedly convex. A skilful thrower can make it describe many varieties of curved paths and even to follow a closed path, by coming back to the thrower himself. Again in the earlier stages of the World War, before the artillery had reached its greatest degree of perfection, it was not infrequent that a shell, with front end conical and base flat, landed hind part before. In other words its axis must have rotated 180° at least once, if indeed it had not made many complete revolutions, end over end, in its flight. And yet its long axis was an axis of symmetry in every respect. Further, while all projectiles from rifled cannon revolve around their long axis during their flight, yet also during the earlier part of it they wobble, that is their point describes not a straight line or slowly bending continuous curve, as does the center of gravity, but a spiral with this line or curve as its axis. Another excellent illustration is the skipping or ricocheting of a flat stone thrown in a certain way against the surface of a pond.

From these well-known illustrations it is quite easy to see that a non-spherical meteor, or at least one not approximating to this shape, will in general rotate about some axis, after encountering the resistance of the air, and may have its course completely changed from an approximate arc of a great circle to a number of curious forms. Among those reported by many observers, and sure to be seen among every thousand or so meteors, are short, sharply curved paths, spirals, great circle paths followed by a meteor which throws off a fragment somewhere which pursues an independent path of different shape, long paths of some slow meteors which show decided curvature from the arc of a great circle, paths pursued by meteors with irregular velocity, meteors which appear, disappear or grow faint, and then finish the latter part of their course with original or greater brightness, etc.

In two special cases small figures (figs. 5 and 6) due to Schiaparelli, easily explain the cases of a meteor actually changing direction and

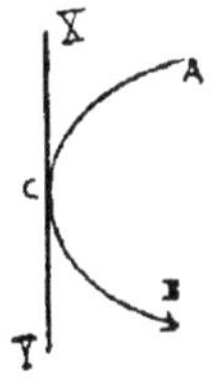

FIG 5

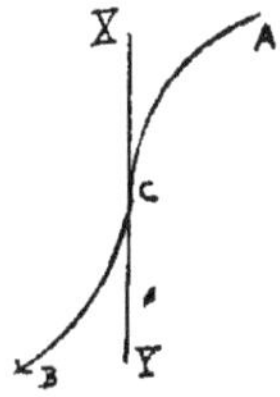

FIG 6

slowing up. Suppose the observer happened to be almost in the plane of motion of the meteor ACB. Let C be the point of tangency of XY which points to observer. Then the meteor would from A to C actually move to left, from C to B to right, or would completely reverse its motion, as seen in projection. In figure 6, when the meteor reaches C it will appear practically stationary, having moved from A to C with constantly decreasing apparent velocity, from C to B with constantly increasing velocity.

The other contributing cause, not mentioned so far except in the summary, is the increasing density of successive strata of the earth's atmosphere. While undoubtedly the density must decrease without sudden jumps, as a general rule, yet the behavior of some meteors certainly calls up in a most convincing way the possibility that they passed suddenly from one density to another. For instance, unexpected slowing up of motion, instant increase in brightness, and a straight path sharply turned aside into the shape of a reap-hook. The general effect of increasing density may be best seen in fireballs, which pursue long paths, frequently coming very low before they burst or disappear. In many such cases the meteor starts as one of perhaps the first or second magnitude only, pursues a long path with apparent fairly regular angular velocity, but continually grows brighter and often changes color as it approaches the earth's surface.

Its increase in brightness often is not a mere effect of coming nearer the observer, for the phenomenon is noted in many cases by persons so situated with regard to the whole path that one end is almost as near to them as the other. The point to be stressed is that many such bodies which begin as mere meteors end as fireballs. A better proof of the intimate connection of the two classes of bodies could not be asked for. It is a pity that we have so little data as to the approximate heights at which the transition, sometimes quite sudden but usually not so, takes place.

A few meteors give the *appearance* of penetrating into our atmosphere and then leaving it, ricocheting as it were Careful researches have given results which seem to prove that for certain widely observed fireballs this thing happened, or at any rate that their end points were much higher above the earth's surface than other parts of their paths. An excellent example of this was a bolide or fireball seen on July 7, 1892. According to the calculations of von Niessl,[4] based upon 34 observations, its perigee was 68 km., but its end point was 158 km. above the earth's surface, its path therein being 100 km. long. The velocity was strongly hyperbolic and its brightness greatly diminished in the latter part of the path. Other similar cases were found by the same investigator.

It will be of interest to see the results of certain observers as to how many paths per thousand deviate from a great circle. Schmidt for instance found 175 for 4068, or 43 per thousand. Zezioli found 104 out of 6853 meteors, or 15 per thousand. The writer out of his first 6200 meteors saw 53 abnormal paths, or 9 per thousand It may be added that care must be taken not to include meteors which are seen just on the edge of the field of vision unless abnormal appearances are certain, because the less well seen a faint meteor is, the more liable one is to think it exhibited some abnormality not seen by another observer looking directly toward it.

[4] *Sitz. d. Kai. Akad. Wien*, **102**, 2A, 265, 1893.

Finally we must state that there is no reason to disbelieve that small bodies, as well as those of planetary size, frequently if not usually rotate about some axis. Then if a meteor thus rotates some of the effects already mentioned will come into play or be accentuated, and it will be the less liable to follow an approximately straight path. This fact is necessary to remember in dealing with all abnormal cases.

Schiaparelli, in §19–21 and *First Note*, taking two empirical laws derived from artillery experiments, calculated how much a meteor's velocity would be slowed up according to how far it penetrated into our atmosphere. This was calculated for the two assumed limiting velocities, 72000 and 16000 meters per second. The meteor was supposed to be spherical, as were the cannon balls with which the experimental data were obtained. The first law called Didion's, is $\phi(u) = 0.026\,(1 + \frac{1}{400}u)\,u^2$, according to Herz,[5] or as Schiaparelli gave it, $R = 0.026\pi r^2 u^2 \left\{1 + \frac{1}{400}u\right\}$, the other law, due to S. Robert, is as follows. $R = 0.03874\pi r^2 u^2 \left\{1 + \left(\frac{u}{676}\right)^2\right\}$ Schiaparelli calculated his table supposedly for a density 3.5, but Herz[5] points out a numerical error and states that the table is really correct for density 2.702. The two tables as given in his *First Note* are shown on page 131.

While the great differences between the corresponding numbers derived from the two laws give us little confidence in their exactness, when thus extrapolated, yet they clearly prove one rather strange thing: a meteor's velocity will be greatly reduced in even those strata where the pressure still is very small. We safely may assume a greater retardation for non-spherical bodies. It again proves to us that all observed velocities must be somewhat smaller than the true velocities. The altitudes corresponding to any of these pressures in Schiaparelli's table may at once be read off from the table, by Humphreys, which appears earlier in this chapter.

Direct observations, though fragmentary, seem to show that in the upper strata the diminution of velocity is smaller but in the lower strata greater than one would expect from the tabular values. Various phenomena which are seen when a fireball is slowed up seem to bear out this view. According to von Niessl[6] the dependence of the

[5] Valentiner, *Handwörterbuch d. Astr.*, 2, 155.
[6] See reference on page 136.

altitude of the terminal point upon the velocity on entrance is to be explained as follows. Under probable assumptions Schiaparelli proved that the entrance velocities of two bodies, both velocities large but very different, will under equal conditions be so diminished that at a certain altitude they have still retained only an equal velocity. This condition is due to the resistance of the air, as may be

u_1	u_0 = 72000 m		u_1	u_0 = 16000 m	
	D $\frac{\mu \sec z}{r p}$	R $\frac{\mu \sec z}{r p}$		D $\frac{\mu \sec z}{r p}$	R $\frac{\mu \sec z}{r p}$
km.	*mm*	*mm*	*km.*	*mm*	*mm*
72	0 00	0 000	16	0 00	0 000
68	0 13	0 002	15	0 81	0 041
64	0 34	0 004	14	1 73	0 091
60	0 55	0 007	13	2 80	0 153
56	0 79	0 010	12	4 04	0 231
52	1 06	0 013	11	5 48	0 330
48	1 37	0 018	10	7 24	0 460
44	1 75	0 024	9	9 47	0 637
40	2 19	0 033	8	12 00	0 886
36	2 74	0 044	7	15 38	1 245
32	3 43	0 059	6	19 85	1 829
28	4 38	0 082	5	26 03	2 703
24	5 47	0 117	4	35 18	4 364
20	7 10	0 174	3	50 04	7 894
16	9 54	0 280	2	78 51	17 561
12	13 57	0 511	1	155 29	61 408
8	21 54	1 166	0 5	280 47	165 988
4	44 72	4 644			
2	88 05	17 842			
1	164 84	61 688			
0 5	290 01	166 268			

seen from the table. Yet von Niessl[0] says it must not be concluded from this that the terminal altitude is independent of the entrance altitude. For the meteor occurring at the greatest will convert more kinetic energy into heat and would therefore be consumed more rapidly.

It is therefore, of importance to learn the heights within the atmosphere at which meteors appear and disappear, and at which trains may be found, not only for the study of these bodies themselves but because it is of great value in the study of the physics of of our air.

In 1864, H. A. Newton collected and published[7] all the cases, 342 in number, of which he could find records. He considered this list fairly complete. Since that time numerous lists of a similar nature have been given out by the British Astronomical Association and may be found in their reports. The extensive work of W. F. Denning is generally included therein. Von Niessl and more recently Hoffmeister have made careful researches of many fireballs seen in Europe. Some of the former's results will be found in the Vienna Academy publications, while other results of both will be found in the *Astronomische Nachrichten*. Numerous smaller lists and isolated cases have been published both in Europe and in America.

For the moment leaving out of consideration telescopic meteors and fireballs, the average meteors appear and disappear, as a rule, within a stratum of our atmosphere whose limits are fairly well known. Yet even within this stratum certain classes of meteors undoubtedly have averages which sharply differ from those of another class. Newton in a paper[8] already reviewed (page 51) gave the following results, rejecting certain obviously erroneous cases

Between 0 and	30 km	39 meteors	Between 150 and	180 km.	57 meteors
30	60	114	180	210	20
60	90	243	210	240	20
90	120	277	240	270	8
120	150	106	270		12

The numbers given are weighted relative numbers, not actual numbers of meteors observed. The centers of paths, not ends, are referred to in the table.

Considering it fairly certain that errors of observation accounted for those less than 30 km. and over 180 km , the average height of the mid-points of the meteors' paths came out 95.55 km. For the Perseids and the Leonids he also found:[9]

	BEGINNING	MIDDLE	END
	km	*km*	*km*
For 39 meteors of 1863 August 10–11	112 4	101 2	90 1
For 78 meteors of 1863 November 13–14	154 9	126 4	97 8

[7] *Am Jour. Sci* (II), **38**, 136, 1864.
[8] *Am. Jour. Sci.*, (II), **39**, 193, 1865.
[9] *Am. Jour. Sci.*, (II), **40**, 250, 1865

S. Newcomb concluded[10] that for 9 Leonids doubly observed on November 14, 1867, the base line being Washington, D. C. to Richmond, Va., the mean height of appearance was 102 miles, of disappearance 47 miles, and mean length of path 67 miles. He added that, as it was certain that their durations were not over one second, he believed these data should be changed, because the initial velocity is only 44 miles/sec. He assumed therefore that 75 miles and 55 miles would be nearer correct for the beginning and end points respectively. It can only be added that neither he nor his co-worker, Harkness, probably was sufficiently trained in plotting meteors to make their results very accurate, as no man, no matter how able, can become a skilful meteor observer without practice

In 1896, W. F. Denning gave a table[11] containing data for 107 objects, observed in recent years. His summary is as follows:

Height of first appearance	73 6 miles = 118 5 km. (106 meteors)
Height of disappearance	45 3 miles = 72 9 km. (107 meteors)
Length of path	62 1 miles = 100 0
Velocity . . .	26 9 miles sec. = 43 3 km.

He stated that this is a promiscuous collection of fireballs and meteors, and that fireballs penetrate lower and exhibit longer flight with a lower velocity. One good observation by himself and Booth, which gave altitudes of 208 miles for appearance and 165 for disappearance, he rejected though the meteor was of first magnitude and observations were quite consistent.

In 1898 he published another very important paper[12] in which he discussed the data then known He concluded that a meteor very rarely appears above 150 miles, and seldom above 130 miles Out of 577 meteors for discussion he found 116 began higher than 100 miles He gave 10 extreme cases, ranging from 483 to 184 miles, but correctly states that the 483 height is probably wrong for reasons he clearly gives. The next highest altitude was 216 miles for a meteor of August 11, 1849, calculated by Heis. His general conclusion is that meteors appear about 76 miles (= 122.4 km.) and disappear at 51 miles (= 82 1 km.), while fireballs penetrate to about 30 miles (= 48.3 km.)

[10] *Am Jour. Sci.* (II), **45**, 233, 1868. Abstract.
[11] *Nature*, **57**, 540, 1898.
[12] *Nature*, **57**, 540, 1898.

Von Niessl gives a number of excellent summaries of altitudes and velocities, forming the most extensive ever compiled, in a paper published by him in 1907 and translated into English by Prof. Cleveland Abbe Sr., and published as No. 16 of Vol. 66, *Smithsonian Miscellaneous Collections*. The translation appeared in 1917. Some of the results are here quoted

NUMBER OF METEORS	HEIGHT OF APPEARANCE	HEIGHT OF DISAPPEARANCE	NOTES
	km	*km*	
109	108 5	86 3	Magns. 1 to 5
121	138 6		Large meteors and me-
213		49 7	teorites
147		60 0	Fireballs
57		31 0	Detonating meteors
16		22 0	Meteorites
49	115 0	88 0	Perseids by Weiss

He adds:

This collection of data shows the influence of the larger masses, especially because of the comparatively slight altitude of the stopping point and therefore because of the deeper penetration into the atmosphere. . . . .Thus it may well be proper to explain these different types of shooting-stars as due to a gradual increase in mass, since larger masses experience a relatively smaller resistance in the atmosphere and thus can penetrate deeper than the smaller masses

The next four tables have to do with limiting altitudes, average geocentric velocities, and average terminal altitudes. The first two refer mostly to meteors proper, the last two mostly to fireballs. In table 1, 78 per cent are meteors, 11 per cent meteors from 1 to 4 times the magnitude of Venus, 11 per cent fireballs of all sizes. In table 2 the same percentages run 74, 6 and 20 per cent respectively. In table 3, however, they run 8, 27 and 65 per cent, all of the latter comparable to the moon or sun and 30 per cent detonating while 7 were accompanied by the fall of a meteorite. In table 4, from older literature, the data refer mostly to large meteors (22 per cent detonating).

He then formed two other tables, which we will call tables 5 and 6, based upon the data used in tables 2 and 3 respectively. This was done to study the effect of the distance of the radiant point

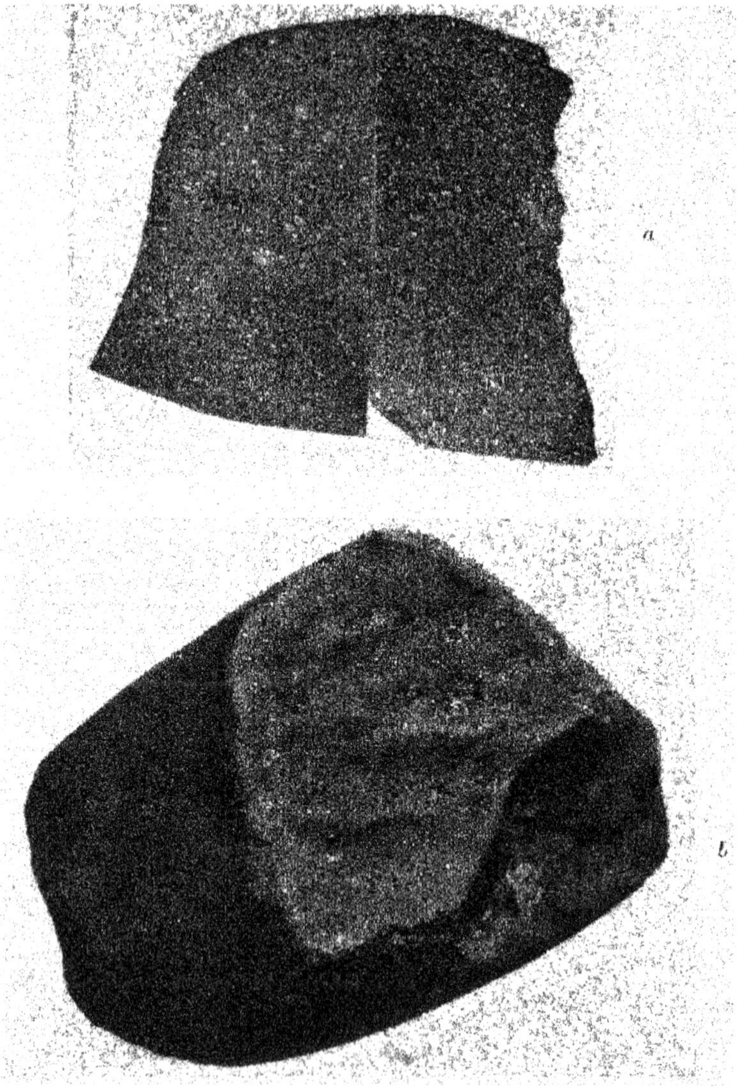

THE SHARPS, VIRGINIA, METEORITE OF APRIL 1, 1921

Photograph loaned by Thomas L Watson Weight 1265 grams (*a*) Sawn surface of the stone, two views (*b*) Complete stone.

from the meteoric apex upon the altitudes at which the various bodies disappeared.

TABLE 1

LIMITING ALTITUDES	NUMBER OF CASES	AVERAGE OBSERVED GEOCENTRIC VELOCITY	AVERAGE TERMINAL ALTITUDE
km.		*km*	*km*
Above 100	23	67 7	106 6
80–100	48	51 5	88 8
60– 80	33	35 5	73 0
30– 60	17	30 1	46 2
Below 30	2	23 2	28 8

TABLE 2

Above 100	10	72 3	112 2
80–100	11	43 0	88 3
60– 80	13	40 5	74 2
50– 60	6	35 4	59 2
Below 50	14	27 4	33 9

TABLE 3

Above 60	12 (1 deton.)	51 8	86 4
50 60	19 (3 deton)	55 0	54 2
30 50	43 (16 deton)	40 6	39 0
Below 30	28 (13 deton)	37 6	24 0

TABLE 4

Above 100	9	76 8	116 8
80-100	9	72 0	89 3
60 - 80	16	49 4	72 5
50 - 60	21	49 1	58 9
30– 50	35	42 7	39 0
Below 30	21	36 5	22 1

All four of these tables show a perfectly regular connection between the geocentric velocity deduced from observations and the altitude of the stopping point since they diminish together It is natural to conclude that a meteor can penetrate into the atmosphere deeper in proportion as it moves with a low velocity . . .

The expected connection between distance from the apex and observed velocity is clearly seen. Table 5 refers to average meteors

TABLE 5

APPARENT ELONGATION OF THE RADIANTS			NUMBER OF CASES	AVERAGE ALTITUDE OF THE TERMINAL POINT
				km.
Between	0° and	40°	12	95 6
	40	70	12	84 5
	70	90	12	61 6
	90	110	10	59 8
	110	180	10	52 1

TABLE 6

Between	0° and	80°	13	54 2
	80	90	9	50 5
	90	100	10	44 5
	100	110	11	40 2
	110	120	7	38 6
	120	150	13	38 6
	150	180	7	36 4

which are destroyed at great altitudes, but table 6 refers mostly to fireballs, which have penetiated deeply into the atmosphere In a sense therefore the tables are supplementary. It is certain that the smaller velocities for the fireballs is partly due to their deeper penetration where the densities are much greater. The final stoppage of the meteor is almost instantaneous, yet observations do show in some cases the effects of retardations in the last part of the path. Reliance on such results, from visual observations, cannot be too great due to the inherent errors of estimating an unexpected and very short phenomenon.

Certain specific cases, in which a fireball's velocity showed decrease, as it penetrated lower, will be given. They are also from the computations of von Niessl.[13]

Fireball of 1905 March 14. End 37 3 km.

LENGTH OF PATH	BETWEEN		VELOCITY
145 km.	87 and	37 km	36 3 km /sec
70	61	37	21 1
51	54	37	14 6

[13] *Sitz d Math Nat. Akad Wien*, **114**, 1477, 1905.

Fireball of 1876 April 9

For 319 km.		80.0 km./sec.
267		76.0
104		29.0

Fireball of Oct. 23, 1889. Began 170 km. End 36.6 km. (For this fireball 14 velocities, corresponding to certain parts of the path, were calculated. They are summed up as below.)

2 parts of path	Over 200 km.	Mean 243 km.	67.2 km./sec.
3 parts of path .	100–200	155	48.2
5 parts of path	40–100	55	25.6
4 parts of path	25–40	32	17.8

Fireball of September 23, 1910[14] For this fireball 15 values are given, the velocities in general showing a decrease (extremes 75.6 to 16.0 km./sec.) for the lower part of the path.

Other examples could be given, but these are quite sufficient to show the great retardations suffered by such bodies as they penetrate comparatively near to the earth's surface. It also shows that the original velocity with which they enter must always be greater than the mean velocity derived by dividing the whole length of path by the whole time of visibility. While von Niessl is unwilling to put any stress upon the actual figures derived, due to the unavoidable and great errors of observation, he considered that the observational data were complete enough to prove the fact quite conclusively.

Having outlined the atmospheric limits within which meteors usually appear and vanish, it will be necessary to give some explanation of how and why a meteor becomes visible and then so suddenly disappears within these limits. Also the fact must be stated that the average meteor [we are not here speaking of or including fireballs] does not appreciably change in brightness during its flight. This is contrary to what commonly might be expected. Further, most visual observations at least do not lend support to any belief in retardation suffered by the average meteor toward the end of its course. However, as the whole phenomenon is over in about 0.5 second, for most meteors, even a trained eye would miss any but a very considerable retardation. In this regard it may be stated that in the mean the latter part of the meteor's flight is nearer the observer, hence a less distance here will fill an equal angle at his eye; that

[14] *Sitz. d. Math. Nat. Akad. Wien*, **121**, 1883, 1912.

is, any retardation is partly compensated by increasing nearness. This effect is a geometrical necessity, but many meteors, due to their relative position, actually are as far or farther off at their end as at their beginning, and the number of cases of meteors growing fainter is small. Nevertheless it must be confessed that it appears incredible, so far as our observations go, that the velocity of a meteor decreases in the ratio of the calculated increase in the density of the air as the meteor penetrates lower.

Some of these questions have recently been treated by E. Öpik in a very important paper.[15] His discussion of probable masses and depths of penetration into the atmosphere are here of special interest. His starting point is the assumption that the energy of the meteor which is converted into ether waves will form a constant fraction of the whole amount. By further assuming that the mass μ varies as the mean zenithal magnitude (i e., the magnitude the meteor would have if in the zenith), and the time of visibility through which the length of the path enters as a function, he was able to work out a table. This was for Perseids only, a meteor of magnitude 2 being used for the unit of both zenithal brightness ι_0 and mass μ.

m_0	ι_0	μ	m_0	ι_0	μ
5 0	0 06	0 029	0 5	3 98	5 9
4 5	0 10	0 052	0 0	6 31	10 6
4 0	0 16	0 095	−0 5	10 0	19 0
3 5	0 25	0 17	−1 0	15 8	34 4
3 0	0 40	0 31	−1 5	25 1	61 9
2 5	0 63	0 55	−2 0	39 8	112 0
2 0	1 00	1 00	−2 5	63 1	201 0
1 5	1 58	1 80	−3 0	100 0	363 0
1 0	2 51	3 25	−3 5	158 0	655 0

Let the radiation due to the solid nucleus be ι_1, the radiation of the gaseous shell around the meteor be ι_2, then the total amount $\iota = \iota_1 + \iota_2$. Therefore $\iota_1 < \iota$, and thus $\iota_1 = \iota$ must be the upper limit of the mass. As to the effective temperature, roughly it is of same order as the sun's. Colors of meteors give some information here as to the probable temperatures However, 3000° to 4000° seem too low an estimate for the Perseids. Öpik's observers thought

[15] *A Statistical Method of Counting Shooting Stars, etc.*, *Pub. Astr Obs.* of *Tartu (Dorpat)*, **25**, No. 1, 1922

the star α Persei was about the average color of the meteors and this is of type F5. Hence it is inferred that the effective temperatures of the meteors is about 7000°. For a height of 100 km., a Perseid of magnitude 2 would have:

TEMPERATURE	DIAMETER	MAXIMUM MASS (DENSITY 4)
	mm	*mgm*
6000°	1 7	10 4
7000	1 3	4 4

He was able to derive for the minimum mass a value of 0 3 mgm., by methods he considered much more reliable, but still assuming 6000° to 7000° as the effective temperature. For temperatures 3000° and 12000° respectively, the minimum mass came out 1.7 and 0.6 mgm. He added that allowing for the rapid increase of emission with increase of temperature it seems probable that a considerable, if not the greatest part of the energy, is lost through radiation. The computed minimum mass will then be the best approximation of the truth. For a 2 magnitude Perseid he believes 1 mgm., to be about the true value. At a temperature of 7000° it means that 37 per cent of the radiation is due to the solid (or fluid) body, 63 per cent to the gaseous particles carried back The above calculations were based on the cosmical velocity of 56 km/sec., and only zenithal magnitudes were used

Having calculated the intensity of evaporation, he states that this quantity determines the hypothetical variation of brightness of the meteor during its flight The quantities of vaporized matter expand, forming the gaseous shell and in consequence of the great area of resistance, are instantly stopped, transferring their kinetic energy into radiation. The next table gives μ = mass, θ = time, J = intensity of evaporation, and v = the corresponding velocities in km./sec.

It is noted what a surprisingly small diminution in the velocity is found. For smaller velocities the results are, however, very different and the slowing up more rapid, while the mass still remaining, when v = 0, is no longer negligible In his general conclusions are found:

(4) . . . . the meteor radiates equally in all directions

(5) . . . . it appears that during its visibility the nucleus of the meteor maintains its initial velocity almost unaltered, the retardation being less

the higher the velocity, the main loss of kinetic energy takes place only after vaporization.

μ	θ	J	V
1 0	0 000	0 000	56 0
0 9	0 105	0 302	
0 8	0 171	0 357	
0 6	0 291	0 384	
0 4	0 411	0 348	
0 2	0 557	0 255	
0 1	0 658	0 175	55 2
0 05	0 738	0 116	
0 01	0 852	0 043	54 4
0 00	1 000	0 000	

Next results from a remarkably important paper,[16] by F. A. Lindemann and G. B. Dobson, entitled *A Theory of Meteors and the Density and Temperature of the Outer Atmosphere to which it leads*, will in part be used. From Denning's data (presumably the latest mean of all good determinations) they state that meteors appear between 160 and 90 km., usually disappear below 120 km. (mostly at 80 km) and have velocities of from 10 to 160 km /sec. As a basis for their work they assume a first magnitude meteor appearing at an altitude of 100 km., ending at 80 km., having travelled a distance of 60 km. at a velocity of 40 km /sec.; in other words, being visible 1.5 seconds, and giving an output of energy at the rate of 4400 h.p. They give the explanation of meteors as follows:

A meteor appears when a cosmic particle of matter moving at a sufficiently high speed relative to the earth, becomes heated by atmospheric friction to such a temperature that it evaporates, and that it disappears when all the matter originally present evaporates. The molecules which distil off of the meteor, as also the molecules of air in its path after collision, are moving at approximately the speed of the meteor and lose their kinetic energy largely in form of radiation, and by collision with other atmospheric molecules.

The meteor moves faster than a sound wave and therefore faster than the air molecules, so on its course it piles up in front an air cushion which is raised to a very high temperature by adiabatic com-

[16] *Proc Royal Soc of London*, **411**; 102, 1923
Jour., R.A.S. of Canada, **17**, 291, 1923.—Abstract by J. Satterlee.

pression. They conclude that most of the energy lost by the meteor goes to accelerating the air, that going to merely compressing the gas being negligible. Some of the heat generated flows from the gas into the meteor, for at high pressure the molecules strike directly upon the cap of compressed gases in front of the solid particle. It is the heat from the gas cap that is directly transferred to the meteor and does the appreciable heating. They give as a further explanation the following:

> The only tenable explanation is that the meteor becomes visible when its surface becomes hot enough to evaporate appreciably, and its fast vapor molecules collide with the air molecules It disappears when it has evaporated practically completely The velocity changes little in the process, also the radiation that comes to us and by which we see the meteor has its source in the vapor; the radiation from the particle being neglible

Among other results they find the mass of the assumed typical meteor to be 6×10^{-3} grams, i.e., if the meteor were made of iron its diameter would be about 0 04 inch. If a meteor had the same speed and duration as mentioned and should appear as bright as the full moon it need only be 1 inch in diameter The reason why meteors appear suddenly at about their average brightness is given as follows: The meteors at first collide so violently with the air molecules met that the latter are broken up and nearly all the energy lost by radiation. It is only when evaporation is considerable and the cloud of vapor slowed up by the air that the meteor becomes bright. Evaporation can only become considerable when a fraction of the heat penetrates into the particle Little time will be needed in the case of a small meteor (They assume that for meteorites the theory would not hold or would need great modification.) By making an equation between the rate of flow of heat to the surface and away into the interior, the density of the air at the place of appearance could be worked out with fair accuracy. By other means they worked out the density at the place of disappearance.

This is not the place to discuss the meteorological aspects of this paper yet some results are so surprising that they merely will be stated. A density at 150 km. altitude is found about 1000 times that formerly calculated. From 60 to 160 km. altitude they calculate a temperature of 300° k, instead of 220° k as formerly held. They definitely conclude against the hydrogen atmosphere, which has so far been usually postulated in the upper strata. The validity of all their results depend upon equations based upon certain assumptions and can be successfully criticised only by a mathematical physicist.[17] It has already been stated that the authors of this paper are now at work in England attempting to secure further and more authentic data by photography of meteors

As for other calculations of the masses of meteors, Harkness in 1867,[18] on three different assumptions (i.e., that the meteor's light is proportional to material consumed in coal gas, Drummond light, and electric light) obtained for a first magnitude meteor 21.6, 0 929, and 0.069 grain respectively or in grams 1.40, 0.06, and 0.0045.

A diametrically opposite point of view is expressed by W. H. Pickering in an article already referred to in the study of stationary radiants. (See page 108.) His statement is that the small mass usually derived from a meteor is based upon "some very doubtful assumption with regard to the amount of energy converted into light by a candle, and the amount of energy similarly converted by a meteor." In *Annals of Harvard College Observatory*, *41*, 40, he deduced that if the intrinsic brilliancy of a meteor of the third magnitude was about that of the incandescent portion of the carbon of an electric arc light, the meteor itself must be 6 or 7 inches in diameter. This intrinsic brilliancy he considered a fair estimate since either iron or stone would volatilize at a lower temperature than the electric arc. If the meteor weighed one gram, its intrinsic brilliancy would be 4000 times that of the incandescent carbons or 40 times that of the sun. He then refers to Trowbridge's article (see page 108) and says it is difficult to believe that a body weighing a few grains could excite a cylinder of space, which may include several cubic miles, in the manner there described. However, if the average meteor were 6 or 7 inches in diameter, then fireballs might be 5 or 6 feet, or sometimes larger.

He then quotes from *Harvard Circular* No. 20 which describes the spectrum of a bright meteor, photographed by chance. The observation is so important that a full description will be taken from the circular. The meteor appeared on July 18, 1897, about 11.00 p m , and was taken with an 8-inch telescope, with objective prism The spectrum consisted of 6 bright lines whose intensity varies on different parts of the photograph, proving the light of the meteor varied.

[17] *Proc Royal Soc of London*, **339**, 103, 1923

[18] *Am. Jour Sci* , II, **45**, 237, 1867.

The approximate wave lengths were 3954, 4121, 4195, 4344, 4636 and 4857, and their estimated intensities were 40, 100, 2, 13, 10, 10 respectively. The first, second, fourth and sixth were probably $H\epsilon$, $H\delta$, $H\gamma$, and $H\beta$, whose wave lengths are 3970, 4101, 4341 and 4862. The fifth line was probably identical with a band at 4633 in stars of fifth type. Of the four hydrogen lines $H\delta$ was most intense. (Note: The description of this spectrum is due E. C. Pickering.) We are necessarily left in doubt whether the hydrogen spectrum was due to gas the meteor had within its mass or to hydrogen already in the atmosphere.

Returning from the digression, W. H. Pickering states that as most of the light came from the bright lines in the spectrum, this indicated that the nucleus was not very incandescent. He further states that: "It can be shown mathematically that on account of the great rarity of atmosphere at these altitudes the pressure in front of a meteor cannot exceed a few atmospheres" He also thinks that temperatures would constantly be kept down by evaporation, so that at ordinary pressures it would be impossible for a high incandescence to exist, and that conduction could play little part, due to the short life of the phenomenon.

Pickering's conclusions then oppose the idea of very small masses for meteors. Unfortunately the question can as yet be solved only by assumptions, and we know so little of the true conditions existing in the upper air. The solution again falls into the domain of the mathematical physicist rather than of the astronomer, and the writer does not venture to say who is right. Certainly a priori larger values would seem more probable, but most men who have attacked the problem find the smaller. Also if the gas cap theory of Lindemann and Dobson, earliest stated in part by Schiaparelli,[19] is true the bright lines could come from the incandescent gas cap, not in general from the nucleus itself. This modification must also be made in the above theory with all the serious consequences which follow.

The other two photographic spectra were secured by S. Blajko[20] at the Moscow Observatory in 1904. The first was on May 11 and referred to a yellow, first magnitude meteor. The second was on August 12 and was due to a Perseid of pure green color and variable

[19] *Sternschnuppen*, § 22

[20] *Astroph Jour*, **26**, 341, 1907.

brightness, being part of the time equal to the first magnitude. The tracks of both meteors were photographed also with another instrument directed toward the same point as the prismatic camera with which the spectra were secured Difficulty was found in getting reliable reference points, but eventually Blajko believed these were determined in fairly satisfactory manner He then identified in the first meteor the H and K calcium lines and a fainter line of calcium, a magnesium and a potassium line. No trace of a continuous spectrum was seen. The meteor brightened suddenly near its end.

For the Perseid meteor, which was less certain due to its position near the edge of the plate, five helium lines were found, and also a thallium line, to which element the bright green color of the meteor was due, in his opinion. Again no continuous spectrum could be seen. The emission spectra of the two meteors were wholly different from one another.

A great deal of visual work on meteors, with hand spectroscopes, was attempted during the last third of the nineteenth century. Such results are inherently too uncertain to be of much value and will not be reviewed. A single exception will be made for an observation of von Konkoly, October 13, 1874,[21] who, with a direct vision spectroscope on a small telescope, observed a meteor train he could trace for 25 minutes with a comet seeker He considered he recognized sodium and magnesium lines, also four bands which on comparison with Geissler tubes he identified as marsh gas.

EXPLOSION OF BOLIDES

It is well known that many fireballs as well as meteorites explode high above the earth's surface. Also that their passage through the air is accompanied by sounds like violent thunder, heard for long distances on either side of the path. In 1919, Ch. Fabry of the University of Marseilles published what appears to be the most correct theory of the whole set of phenomena so far presented.[22] His results follow in an abbreviated form.

He states that there are two distinct phenomena, the bursting of the bolide and the violent noise produced by its rapid movement, identical with the phenomena known to the artillerists as the shock

[21] *Astr. Reg.*, **12**, 3, 1874.
[22] *Bul. Soc. Astr. de France*, **33**, 448, 1919.

wave (l'onde de choc). The bolide first has an enormous pressure upon its anterior face due to the resistance of the air. While exact calculation is difficult, for a bolide 20 cm. in diameter this must at least be a dozen tons, quite sufficient to break it. At the same time, rushing through the air warms the surface of the bolide and unequal expansion of different parts assures rupture. Finally if the bolide contains any trace of volatile substance this can be reduced to gas and will make it shine. It is added that the last reason is a little hazardous if the bolide is thought of as having just crossed empty space, exposed to the rays of the sun. However that may be, the bolide bursts with an explosion capable of waking people for leagues around. We cannot seriously suppose that the bolide contains actual explosives. And we know that its mere bursting from crushing or unequal expansion could not produce the tremendous sound frequently reported

Fabry adds that this formidable noise has no association with the actual rupture, it is simply due to its rapid motion through the air producing a shock wave. This phenonemon is known as a ballistic wave. It consists of an apparent detonation which is produced without any explosion when a projectile moves with great velocity through the air The necessary condition is that its speed must exceed that of the propagation of sound It is further necessary that the observer be properly situated with respect to the trajectory, thus persons placed behind would not hear it, but those beneath would hear it best If the trajectory is vertical, the wave of shock would be perceived all around the point of fall, but if oblique some sectors would receive it, some would not, even though actually nearer to the trajectory itself Fabry therefore concluded that the actual explosion does not produce the noise as we might have the latter without the former.

The writer believes that the above theory correctly explains the sound phenonema heard so frequently on the passage of a bolide, but is not prepared to admit that the final explosion itself does not produce a considerable sound, certainly audible near the point of explosion. It seems for instance that the great bolide of May 11, 1922, which ended near Amelia C. H., Va., and of which he has collected numerous observations, certainly exploded *twice*, just about the end of its path, the noise of the explosions coming, with a short interval between them, to an observer near by, who saw the fragments thrown off. Also the intervals of time between when the final explosion was seen and heard, as observed by several others, helped to fix very exactly the point over which the explosion occurred, the azimuths of course being taken into proper consideration. Similar cases published by others could be mentioned. It seems necessary therefore to admit the actual explosion sound, corresponding to the event itself, but not at all necessarily one which would produce the general sound phenonema due to the bolide's passage over other places. With this modification Fabry's hypothesis seems to give an excellent explanation of what actually takes place It is scarcely necessary to add that for bolides which do not explode, and for meteorites which arrive in a single unbroken mass, Fabry's hypothesis needs no such modification.

It should in fairness be stated that a somewhat similar theory had already been advanced by L. Fletcher in his valuable book *An Introduction to the Study of Meteorites*, page 30, 1914. In the same book will be found a brief but excellent treatment of the whole subject of meteorites and their connections with meteors.

CHAPTER XIII

Meteor Trains

Even the most casual observer has noticed that some meteors leave a train or luminous streak behind them, while others seem to leave nothing. In most cases this train is as ephemeral as the meteor itself, but a few hours of observation generally will show one or more that last several seconds, and years of such work will furnish a few that last several minutes. We have on excellent authority that, if a telescope is at hand and is turned upon a meteor train, its visibility is greatly prolonged, so that one which to the unaided eye would disappear in ten or twenty seconds might thus be seen two or three minutes In extreme cases even to the naked eye they have remained visible over half an hour. The smoke cloud left by the explosion of a bolide in daylight sometimes remains visible even longer.

Fairly accurate observations, made simultaneously at two or more places upon remarkable trains, have proved that they were several miles in length and certainly occasionally one or more miles in diameter. We thus know that they seem to fill several cubic miles of space with their faint glow. As again we know positively that the bodies which produced them are all very small, no matter whose estimate we accept, and usually could not on their passage have filled one millionth part of the space so illuminated, it long has been a mystery how such trains were produced. Or, if produced, how they could remain luminous in the (supposed) intensely cold and tenuous atmosphere at that altitude. The chance of irradiation playing an important part in their apparent size, as it does for the fireball or meteor itself, is wholly untenable because telescopic observations, made at leisure, confirm their dimensions. Also the trains are not very brilliant per unit area hence irradiation would not be a large factor.

The explanation that appeared simplest at first sight and that often has been given was that the trains consisted of the incandescent material and glowing gas left by the passing body; in the case of a meteor most of the material going thus to form the train, and for a fireball at least that on the surface. That this explanation is false is obvious, first because the tiny particles which might be assumed to be left, as well as the heated gas, would cool in a second of time or less; second because not enough of them possibly could be thrown off nor enough gas be heated to fill so great a space. It might be added that certain meteors, usually with slow angular motion, do leave a train of sparks behind them, but this is a different phenomenon and each tiny spark is a unit, which quickly ceases to glow, goes out as would a white hot coal of its size. No observer need mix this phenomenon with the usual trains, particularly the long enduring ones.

It so happens that the labors of one man, the late C. C. Trowbridge, of Columbia University, have done more for the study of meteor trains than all others combined, hence his work alone will be reviewed here. He was not himself an observer, hence depended upon the careful data of astronomers, among whom we may mention Barnard, Denning, A Herschel, and Newton, as well as many others. Nevertheless to Trowbridge's zeal in collecting and studying the records obtained by these men is due the best theory of what produces this mysterious train.

These data were tolerably extensive, for in 1907[1] he had collected reports of thirty-seven trains which remained visible from over five to forty minutes each; fifty-three in all over one minute with an average duration of 14.8 minutes. His conclusions are as follows:

1 The meteor trains are self luminous gas clouds combined with very minute meteoric dust particles, the latter in daylight reflecting light like any ordinary clouds.

2 The height of meteor trains seen at night appears to be at a definite altitude indicating that the phosphorescence is dependent on the gas pressure where the trains are found

3 The diffusion of the trains is gas diffusion and its rate depends on the temperature of the atmosphere, and probably on the initial intensity of the train

4. Many meteor trains appear to be tubular in form, that is the luminosity is greater near the border

5. Experiments have been made by the writer which give the law for the rate of decay of the phosphorescence of the air at very low pressure, and these experiments explain the long duration of the meteor trains, on the hypothesis that it is a phosphorescence which decays according to the same law

---

[1] *Astroph. Jour.*, **26**, 95, 1907.

6. Statistics on the color of trains show that, excluding those illuminated by sunlight, trains are as a rule green or yellow fading to white, colors which are typical of the phosphorescence of the air

Table 1, giving altitudes of trains, also is copied.

NUMBER	ALTITUDE LIMITS OF VISIBLE TRACK OF NUCLEUS—MILES		ALTITUDE LIMITS OF TRAIN—MILES		MEAN ALTITUDE OF TRAIN—MILES	ALTITUDE COMPUTATIONS MADE BY
10					50	A S Herschel
12	100	53	59	53	56	A and J Thomson
26					54	A S Herschel
29*	120	60	65	60	63	H A Newton
44	68	49	59	49	54	H A Newton
46					51	H A Newton
47	65	52			58	H A Newton
52†					59	H A. Newton
80	90	30	58	50	54	W F Denning
120	78	47	59	47	53	W F Denning
121	65	37	57	45	51	W F Denning
125‡	65	28			45	W F Denning
129	90	41			54	A S Herschel and Greg
Mean	82 8	44 1	59 5	50 7	54 0	

* Over 60 miles—train five miles long, carefully calculated
† Altitude of lower part of train
‡ Short train ½° long at 45 miles

The average mean altitude of 54 miles = 87 km. seems excellently well determined, and the upper and lower limits of 60 and 50 miles scarcely less so. It further is seen that the altitude at which the nucleus itself became visible seems to have no effect upon the position of the train, in some cases 40 to 50 miles having been travelled before its production. Also in some cases the nucleus continued to move on some distance after the train had ceased to form. That a train is often visible for only a part of a meteor's course is a phenomenon well known to every observer, as well as the fact that the nucleus sometimes goes farther without the train following to the very end. This fact was stated in 1866–1867 by Greg, Glover, and Goulier independently.

As for meteor trains seen in daylight, according to Trowbridge[2] they seldom occur above 40 miles and sometimes as low as 25 miles.

[2] *Pop Sci Monthly*, **79**, 191, 1911

They appear as if composed of thin smoke. This, in connection with the much higher altitudes of the night trains, seems to indicate that in the upper levels of the atmosphere the glow does not arise mainly from light reflected from fine meteoric dust, but is a luminosity of the gas in the meteor's train. Attention is called to the Leonids leaving greenish trains, the Perseids yellowish, and most daylight trains (i.e., smoke) appearing reddish—this last merely reflected sunlight. His general idea is that the night trains, which for long durations always expand, are due to gas diffusion but are tubular in form. Hence when seen from the outside the two sides of the relatively empty tube would appear brighter than the middle, which is actually confirmed by observations. The conclusion, in connection with laboratory experiments, is that the barometric pressure at these heights, i.e., 50 to 60 miles altitude, is not far from 0.2 mm.

In his second paper mentioned he elaborates what occurs on the passage of a meteor as follows:

> It has been shown . . . that when a body is very hot an immense number of negatively charged corpuscles or ions are given forth by the body. Air containing free ions becomes a conductor of electricity, hence we have in a meteor rushing through the atmosphere a condition extremely like a very long electrical discharge tube containing gas at low pressure. . . . . The burning meteor . . . must form a column of highly ionized air . . . Moreover at a certain altitude, corresponding to a pressure 0 2 mm . . . . or about one two-thousandth to about one four-thousandth of one atmosphere pressure, the conditions are precisely right for the formation of phosphorescence in the meteor track . When the meteor nucleus has been consumed, all that remains visible in the dark sky is the body of phosphorescent gas in the part of the track where the gas pressure conditions were correct for the formation of the persistent glow

He adds that it is not certain that electrical discharges take place in the meteor's track, and they may not even be essential for the formation of the phosphorescence. The air heated and ionized by the rapidly moving, burning meteor may readily suffer chemical or physical changes in its composition which on gradually reverting to its original state gives out such a glow. Both papers quoted, as well as others by the author, contain much valuable detail and excellent drawings of long enduring meteor trains, as well as a fuller explanation of the theory than can be given in this brief review. To date they may be considered classical in this subject.

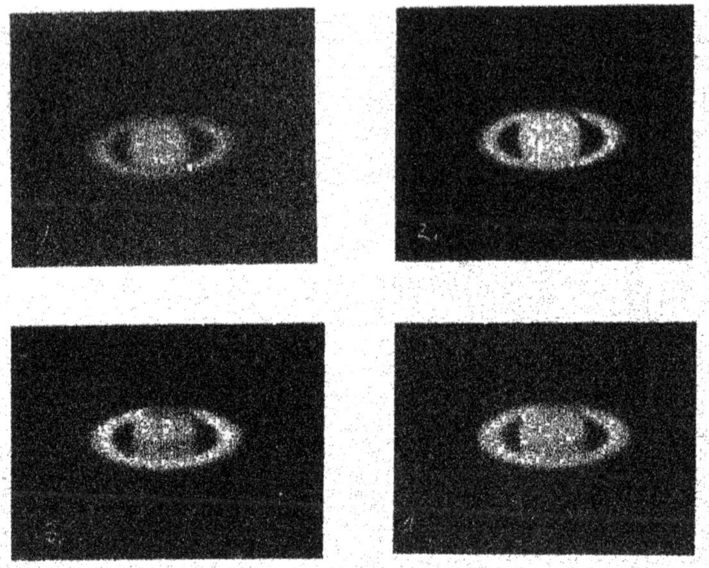

FOUR MONOCHROMATIC PHOTOGRAPHS OF SATURN BY R W. WOOD

The Rings are composed of meteoric matter

1 Infra-red
2 Yellow
3 Violet
4 Ultra-violet

The study of meteors in our atmosphere has so far been limited to those visible to the naked eye. Yet any astronomer knows that in a telescope meteors are observed far too faint to be seen without optical aid. The faintness of these meteors is limited only by the size of the telescope. For instance with the McCormick 66 cm. refractor, one easily can see 14 magnitude meteors, and under the best conditions 15 magnitude—in other words within one magnitude of the telescope's stellar limit.

Now if we assume that the average meteor, seen without a telescope, moves 10° in 0.5 second, which is not far from correct, its angular motion is therefore 20° per second. If our telescope is a large one with a moderate power the field will be anywhere from 5′ to 20′ in diameter. As 20′ is $\frac{1}{60}$ of 20°, then such a meteor would pass through the field in $\frac{1}{60}$ second. However one might work a lifetime using high powers on a telescope and never see a meteor as bright as the sixth magnitude cross the field of view. With fields such as are available for comet seekers or smaller telescopes in general, numerous faint meteors may be seen, and occasionally a brighter one Only a few observers have published real lists of telescopic meteors. Among these few Schmidt made 146 such observations in 1844–1851;[3] Winnecke in 1864[4] made 68; and E. P. Mason made about 30 observations at the 1839 maximum of the Perseids.[5] More recently the American Meteor Society, with the aid of the American Association for Variable Star Observers, has taken up the task of collecting such observations, and published a list of 152 without full details.[6] Many more unpublished observations are in its files. The first single observation of a telescopic meteor apparently was made by J. H. Schroeter[7] on June 28, 1795. Argelander considered those of the 7, 8 and 9 magnitudes as being 40 times more numerous than those visible to the unaided eye. Mason's observations in 1839 are so interesting that they are quoted in part:

> During four or five evenings about August 9, between 20 and 30 meteors passed . . . . about 20 on the 9th and 10th. . . . . Their apparent

---

[3] *Handworterbuch der Astr.*, **2**, 124. Abstract.

[4] *Wochenschrift f. Astr*, 377, 1861

[5] *Trans, Am. Philos. Soc.*, *N. S.*, **8**, 122, 1843.

[6] *M* 3.

[7] *Handworterbuch der Astr.*, **2**, 120, 1898.

brightness and velocity, as magnified . . . were, on an average, about the same, or rather less than those seen by the naked eye. . . . . They were of a sensible size, more so than that of ordinary meteors of the same absolute brightness. On an average they were about half or one-third the diameter of Jupiter . . . their outline was somewhat indefinite, like a star out of focus . . . Unless there is reason for a great difference between the absolute velocities of the more distant and nearer of these bodies, the telescopic meteors seen could not have been much less than 80 times as far above the earth as those seen by the naked eye, which (are about) 50 or 60 miles. This latter quantity, multiplied by 80 or the magnifying power of the telescope, indicates a probable elevation of at least 4000 miles. . . . .

Kleiber calculated that a person could cover an area of the sky 80° in diameter. Schmidt's telescope had a field of about 3°. Therefore the relative areas are roughly $\left(\frac{3}{80}\right)^2$ or more exactly 1:683. The diameter of the field of Winnecke's instrument seems also to have been 3°. Newton[8] who assumed this field to have been only 53′ (that is less than one-third of the diameter that Klein gives[9]) calculated on other assumptions that 1582 were visible per hour in the telescope if the whole sky could be covered relative to the 8 per hour seen on an average by one person This was based upon Winnecke seeing 45 with his telescope, while Pape observing with him, but without optical aid, saw 312, between July 24, to August 3, 1854 in 32 hours of work.

It is curious that Newton here overlooked the obvious fact that if we consider Winnecke to have had his eye actually at his telescope all the time, in that restricted field no meteor would have been missed. On the contrary the 8 per hour, usually found by one person without a telescope refers to only a small fraction of the meteors > 6 magnitude that can be counted per hour in the whole sky. This would greatly change his results. The writer deduces 45:312, and area of zone 3° in diameter: area zone 80° in diameter; that is in 683 times the area that Pape actually saw 7 times as many meteors per hour. Or more exactly the ratio is 1:98. But an area R = 40° is to that of the hemisphere as 1:4.27, therefore 421 telescopic meteors would have been seen per hour. Using Newton's value of 53′, the number would be increased by $\left(\frac{180}{53}\right)^2 = 12\pm$, or $421 \times 12 = 5052$. Again as

[8] *Am Jour Sci* (II), **39**, 200, 1865

[9] *Handbuch der Astr. Himmelsber*, 285, 1901.

Newton used 8 per hour instead of $\frac{312}{32} = 10\pm$, this would be reduced by $5052 \times \frac{8}{10} = 4042$.

The next article of importance was by Professor Schafaric at Prague, which appeared in 1885.[10] He speaks of having seen hundreds of such meteors in his 6½ inch reflector, field 45′. On August 30, 1880, 9 to 15 hours, he says he thought 50 to 100 passed; on the next night 20 at least. He divides all such objects into four classes, which will be quoted as no other such classification is known to the writer:

1. Well defined star like objects of very small diameter, round sometimes with smoky luminous tracks of cometary aspect, i e , widening as they recede from the meteor

2 Large luminous bodies of some minutes in diameter, round or ovoid, sometimes well defined, ordinarily diffused . . .

3 Well defined disks of very perceptible diameter, brighter at edge than center. . . .

4. Faint diffused nebulous masses of irregular shape, considerable size and different colors

He gives numerous examples from his very interesting observations, and explanations for the physical constitution of each class. He evidently had very great fortune in detecting telescopic meteors.

An even more interesting article appeared in 1914 by W. F. Denning[11] He states that he has "seen more than 1000 meteors in telescopes." For instance in the years 1881–1896, in 727 hours of comet seeking, he saw 635 meteors in his 10-inch reflector. Using similar methods of deduction to those of Mason, for 4 objects which were examples of a class and not rarities, he inferred altitudes of from 1260 to 1820 miles. He added that often very swift meteors also are seen, which obviously are near, and that, as no authentic case of meteors visible to the unaided eye and having an altitude of over 200 miles is known, therefore the more distant meteors possibly are different in material, nature and motion and depend for their light upon conditions widely dissimilar to those of ordinary meteors. In the same volume of the *Observatory*, page 417–419, he gives a short

[10] *Astr. Reg* , **23,** 205, 1885.
[11] *Observatory*, **37,** 211, 1914.

article quite effectively proving that the so-called daylight telescopic meteors usually are terrestrial in origin and not true meteors at all.

In 1899, T. J. J. See[12] gave a note on the telescopic meteors seen in 1896–1898 with the Lowell 24-inch refractor. He said: "A careful consideration of our whole experience, based partly on counts, indicated . . . . not less than five meteors per night would cross the field (6′ in diameter)." From this he deduces a daily number of 12×10^8 telescopic meteors for the whole earth.

Unfortunately both the numbers of Schafarak and See are based partly upon mere impressions, which are likely to lead to erroneous results. For instance a well known member of the A. A. V. S. O. reported to the writer that he had seen only one telescopic meteor during the whole year's observing, but added "that he seemed to remember, before he began to record them, seeing one or more almost every night."

From those seen and carefully recorded by the writer, he can add that, with some exceptions, as seen in the telescope they do not appear to move with appreciably faster angular motion nor to be dissimilar to those seen without such aid. The usual—and obvious—reason given is that we see only faint meteors so near their radiants that the foreshortening of their paths makes their angular apparent motion moderate. Nevertheless, while unable to contradict this, it certainly appears to an observer that many meteors seen in a telescope are not near their radiants. For instance the writer is quite sure that he observed a 10 magnitude Orionid with the 66 cm. telescope, 20° from the radiant, on October 22, 1922. Thus there seems no escape from one of two opposite conclusions: either all such meteors are at the usual heights but very near their radiants, or the atmosphere is several times higher than we usually suppose—as was first indicated by Mason, and again by Denning, in observations quoted.

As an alternative the very instructive suggestion of Denning, already mentioned, that perhaps meteors become visible at very great altitudes through unknown causes certainly deserves careful consideration. This, if true, would no longer necessitate our considering the atmosphere to be of very great thickness. Again the fact that auroral streamers appear up to 500 km. altitude is not absolute evidence that the atmosphere, as we usually consider it, ex-

[12] *Astr. Nach*, **151**, 297, 1899

tends to such a height. But the discussion of this point lies beyond the scope of this book.

One way to settle the question is to place, for instance at the Perseid maximum, two telescopes with wide fields of view, about one or two miles apart, and have two careful observers train them constantly upon the region of the radiant. By means of the B. D. maps, enlarged somewhat, or photographs of the region, any telescopic meteor could be accurately plotted, and its height then derived in the usual way. As a most important observation its carrying out is strongly urged upon others interested in the subject.

A very unique observation made by L. J. Wilson[13] at Nashville, Tenn., a member of the A. M. S., will be here reprinted:

> On May 17, 1911, during an observation of Jupiter, a meteor fully as bright as that planet, passed 35′ above it, moving E to W, and the train quickly disappeared. May 17 at 10 40 C. S. T., was one night when seeing was perfect, and as soon as I looked through the telescope after the passage of the meteor, the air was in such violent motion that only the coarsest features of Jupiter could be seen. The air vibrations were at first very rapid, less so as the seeing improved, until finally after a few ripples the seeing became normal. The duration of the disturbance was between 4 and 5 minutes. Color of meteor that of Jupiter, path 30° long.

In view of what has been said it appears that a more accurate study of meteors and their phenomena in our atmosphere will enable us to gain information of great value to many different branches of science. But not until automatic means of recording their velocities, used in connection with photography, are employed, can data of the requisite accuracy be hoped for. Even then, unless many stations are so equipped, the gathering of the data will be relatively slow. Nevertheless the sooner it is undertaken, the sooner will the solutions of these problems be within reach.

[13] *M* 2, 475, 1914.

CHAPTER XIV

Computation of Real Heights of Meteors

Of the various methods which have been published the most convenient and accurate seems to be that which originally appeared in *Contributions of the Lick Observatory*, No. 5, 1895.[1] It is fully reproduced here with the permission of Director W. W. Campbell. The method itself was developed by J. M Schaeberle and first applied by him to observations made by several of his colleagues and himself during the Perseid maximum of 1894 The general discussion of the method and the numerical example given on page 162 are, however, due to the writer.

To determine the actual position of the beginning and ending points of a meteor's visible path with regard to the earth's surface it is necessary that these points in the sky be observed from at least two stations, whose geographical coordinates and distance apart are known. As it frequently is difficult to be absolutely certain that the same meteor has been observed from both stations the following two conditions must be fulfilled. First the times at which the object is seen from both stations must correspond to the same absolute instant. Second the points of apparition and the points of disappearance, as seen from both stations, must respectively lie in two planes which also contain both stations

The first condition is found by merely comparing the corrected times of observation at both stations. The second condition can be made clear by figure 7. In this let S_1 and S_2 be the two stations. Let P be the point in our atmosphere at which the meteor actually appears or disappears. Then the point P and the line S_1S_2 define a plane which cuts the celestial sphere in a great circle BX_1X_2A. Obviously X_1 is the projection of P upon the sphere as seen from S_1, and X_2 the projection as seen from S_2, while A is the point in which the line S_1S_2 pierces the sphere as seen from S_1. Therefore, that the second condition may be fulfilled, the points X_2, X_1, and A must lie upon the same great circle, the equation of which is given in (4).

[1] *Bul Soc Astr de France*, **17**, 322, 1903, contains a French translation of the original article and also a simplified method, based thereon, by H. Chrétien

To develop the equations, the following quantities are to be defined.

R_1 = the radius of the earth at S_1,

ϕ_1' = the geocentric latitude at S_1,

θ_1 = the local sidereal time of the observation at S_1;

R_2, ϕ_2', and θ_2 correspond to the same quantities for S_2.

Referring the two stations to the earth's center as origin, the plane of the earth's equator being chosen as the XY plane, with the +X axis directed to the vernal equinox, the +Y axis to a point with $\alpha = 90°$, $\delta = 0°$, and the +Z axis to the north celestial pole; then

$$(1)\begin{cases} X_1 = R_1 \cos \phi_1' \cos \theta_1, \\ Y_1 = R_1 \cos \phi_1' \sin \theta_1, \\ Z_1 = R_1 \sin \phi_1' \end{cases} \qquad (2)\begin{cases} X_2 = R_2 \cos \phi_2' \cos \theta_2, \\ Y_2 = R_2 \cos \phi_2' \sin \theta_2, \\ Z_2 = R_2 \sin \phi_2'; \end{cases}$$

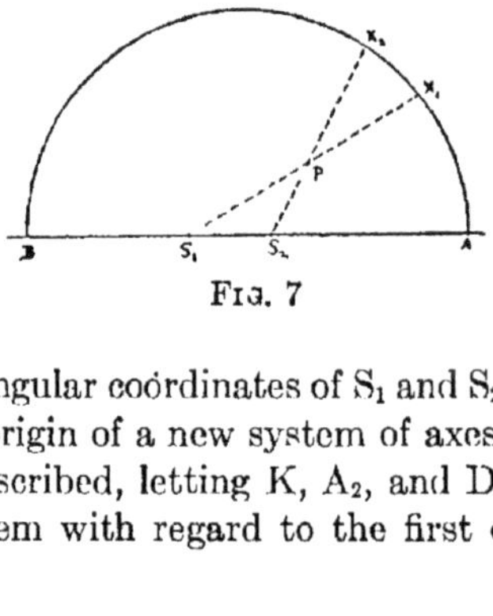

FIG. 7

represent the rectangular coördinates of S_1 and S_2 respectively. Now let S_1 become the origin of a new system of axes, parallel to those of the system just described, letting K, A_2, and D_2 be the coordinates of the second system with regard to the first on this new system. Then

$$(3)\begin{cases} K \cos D_2 \cos A_2 = X_2 - X_1, \\ K \cos D_2 \sin A_2 = Y_2 - Y_1, \\ K \sin D_2 \quad\quad = Z_2 - Z_1. \end{cases}$$

If we now substitute the proper values of (1) and (2) on the right hand side of (3) we can at once solve for K, A_2, and D_2, having three equations and only three unknowns. Now let the coördinates of X_1 be α_1, δ_1, those of X_2 be α_2, δ_2, and those of A be A_2, D_2. Then we must have the three points (α_1, δ_1), (α_2, δ_2) and (A_2, D_2) lying upon a great circle. Also let I be the inclination of this circle to the celestial equator and N one of the points in which they cut. If now we draw arcs of great circles from X_1, X_2, and A perpendicular to the celestial equator we construct three right spherical triangles. For these we can at once find that

$$\begin{cases} \tan D_2 = \pm \sin (N-A_2) \tan I, \\ \tan \delta_1 = \pm \sin (N-\alpha_1) \tan I, \\ \tan \delta_2 = \pm \sin (N-\alpha_2) \tan I, \end{cases}$$

where the same sign on the right side must be used for all three equations. Eliminating N and I the final equation is derived:

$$(4) \quad \tan D_2 \sin (\alpha_2-\alpha_1) - \tan \delta_1 \sin (\alpha_2-A_2) + \tan \delta_2 \sin (\alpha_1-A_2) = 0.$$

While this is the exact condition in practice it can at best be only approximated. The speediest procedure is to plot the three points upon a celestial globe and see whether a stretched string can be made to nearly touch them all. If a globe is not at hand only experience can teach just how great a deviation from zero in the equation can be permitted and still there be certainty or great probability of (α_1, δ_1) and (α_2, δ_2) actually corresponding to the same point P.

If ρ_1 and ρ_2 denote the distances of P to S_1 and S_2 respectively, the rectangular equatorial coordinates of P referred to the first station as origin will be

$$(5) \begin{cases} x_1 = \rho_1 \cos \delta_1 \cos \alpha_1, \\ y_1 = \rho_1 \cos \delta_1 \sin \alpha_1, \\ z_1 = \rho_1 \sin \delta_1, \end{cases} \qquad (6) \begin{cases} x_2 = \rho_2 \cos \delta_2 \cos \alpha_2, \\ y_2 = \rho_2 \cos \delta_2 \sin \alpha_2, \\ z_2 = \rho_2 \sin \delta_2 \end{cases}$$

Now since the differences between the two sets of coördinates for a given meteor must evidently always equal the differences between the coordinates of the two stations, then

$$(7) \begin{cases} K \cos D_2 \cos A_2 = \rho_2 \cos \delta_2 \cos \alpha_2 - \rho_1 \cos \delta_1 \cos \alpha_1, \\ K \cos D_2 \sin A_2 = \rho_2 \cos \delta_2 \sin \alpha_2 - \rho_1 \cos \delta_1 \sin \alpha_1, \\ K \sin D_2 \qquad\quad = \rho_2 \sin \delta_2 \qquad\quad - \rho_1 \sin \delta_1 \end{cases}$$

Multiply the first of (7) by $\sin \alpha_2$, the second by $\cos \alpha_2$ and equate, then

$$(8) \qquad \rho_1 = K \cos D_2 \frac{\sin (\alpha_2 - A_2)}{\cos \delta_1 \sin (\alpha_2 - \alpha_1)}$$

In like manner

$$(9) \qquad \rho_2 = K \cos D_2 \frac{\sin (\alpha_1 - A_2)}{\cos \delta_2 \sin (\alpha_2 - \alpha_1)}$$

Now let (α, δ) be the geocentric coördinates of P, then the geocentric equatorial coördinates will be

$$(10)\quad \begin{cases} x = R \cos \delta \cos \alpha, \\ y = R \cos \delta \sin \alpha, \\ z = R \sin \delta. \end{cases}$$

As obviously $x = X_1 + x_1$, $y = Y_1 + y_1$, and $z = Z_1 + z_1$, then

$$(11)\quad \begin{cases} R \cos \delta \cos \alpha = R_1 \cos \phi_1' \cos \theta_1 + \rho_1 \cos \delta_1 \cos \alpha_1 = \\ \qquad\qquad R_2 \cos \phi_2' \cos \theta_2 + \rho_2 \cos \delta_2 \cos \alpha_2, \\ R \cos \delta \sin \alpha = R_1 \cos \phi_1' \sin \theta_1 + \rho_1 \cos \delta_1 \sin \alpha_2 = \\ \qquad\qquad R_2 \cos \phi_2' \sin \theta_2 + \rho_2 \cos \delta_2 \sin \alpha_2, \\ R \sin \delta = R_1 \sin \phi' + \rho_1 \sin \delta_1 = \\ \qquad\qquad R_2 \sin \varphi'_2 + \rho_2 \sin \delta_2. \end{cases}$$

From these equations (11) R, α, and δ can be found in duplicate, the last term of each separate equation serving as a full check upon the middle term. (See example.)

The height of the meteor above sea-level is given by

$$(12)\qquad h = (R - R') \cos (\phi' - \phi)$$

in which R is the radius vector of the earth corresponding to the geocentric declination δ or the equivalent latitude ϕ', the geographic latitude being ϕ. But as $(\varphi' - \phi)$ never exceeds 12′, cos 12′ = 1 for all practical purposes, and we can write $h = R - R'$. This gives h in terms of whatever units we have expressed R and R′.

Finally to obtain the terrestrial longitude of P we have

$$(13)\qquad \lambda = \lambda_1 + (\theta_1 - \alpha) = \lambda_2 + (\theta_2 - \alpha)$$

The point (λ, ϕ) on the earth's surface is that point which has P for its zenith.

It is obvious that when the position (λ, ϕ) and height h of P_b and P_e are determined, P_b designating the beginning point, P_e the end point, then at once a most simple solution gives the inclination to the earth's surface, and the azimuth of the vertical plane containing P_bP_e. Also the length of path P_bP_e can be determined and the velocity, which latter equals $\frac{P_bP_e}{t}$ where t is the number of seconds the meteor was visible.

Special case. In equations (8) and (9) it can be seen that if $\alpha_2 - A_2$, $\alpha_2 - \alpha_1$, or $\alpha_1 - A_2$ is near 90° or 270°, or if δ_1 or δ_2 is near 0°, then (8) and (9) may become nearly indeterminate. In any of these cases we may obtain longer but accurate expressions but ρ_1 and ρ_2. To do this multiply the three equations of (7) in order by $\sin \delta_2 \cos \alpha_2$, $-\sin \delta_2 \sin \alpha_2$, and $-\cos \delta_2 \cos 2\alpha_2$ respectively. Combine these equations to eliminate ρ_2. We then obtain

$$(13) \qquad \rho_1 = K \frac{\sin \delta_2 \cos D_2 \cos (\alpha_2 + A_2) - \cos \delta_2 \sin D_2 \cos 2\alpha_2}{\sin \delta_2 \cos \delta_1 \cos (\alpha_2 + \alpha_1) - \cos \delta_2 \sin \delta_1 \cos 2\alpha_2}$$

Multiplying the same equations again by $\sin \delta_1 \cos \alpha_1$, $-\sin \delta_1 \sin \alpha$, and $-\cos \delta_1 \cos 2\alpha$, and eliminating ρ_1 we obtain

$$(14) \qquad \rho_2 = K \frac{\sin \delta_1 \cos D_2 \cos (\alpha_1 + A_2) - \cos \delta_1 \sin D_2 \cos 2\alpha_1}{-\sin \delta_1 \cos \delta_2 \cos (\alpha_2 + \alpha_1) + \cos \delta_1 \sin \delta_2 \cos 2\alpha_1}$$

A great advantage of this general method of procedure is that in (11) the derived values of R, δ, and α are completely checked by using the different sets of equations given in the second and third columns. In the numerical example on page 162 it will be found that the two values of x and y check almost perfectly, the only difference being in the values of z. The nearer the values of z coincide, the more accurate the observations. When both are given, as in this case, one can determine at a glance by their agreement—or lack of it—just what degree of accuracy the observations attained. It is hard to set any exact numerical rule but in general, if the values of the end points agree within about 5 km. and those of the beginning points within about 8 km., the results would be fairly good for visual work. It is a matter of great regret that the most extensive lists of real heights, as for instance those of the British Astronomical Association, frequently do not give the check computations nor indeed any way of telling how accurate any given case may be, except perhaps a note by the computer saying "good," "uncertain," or some similar remark.

For the average case of a meteor seen by two observers no method so good as the above is known. It should be added that unless the two stations are at least 20 miles apart the errors of observation generally will invalidate the results. When the base-line is over 100 miles long it becomes more difficult to identify the meteor with certainty, particularly during the brighter showers of the year. A base-line of 40 to 60 miles therefore is recommended.

July 23, 15h 27m—18h 2m G M T. July 29, 16h 5m—18h 5m G M T

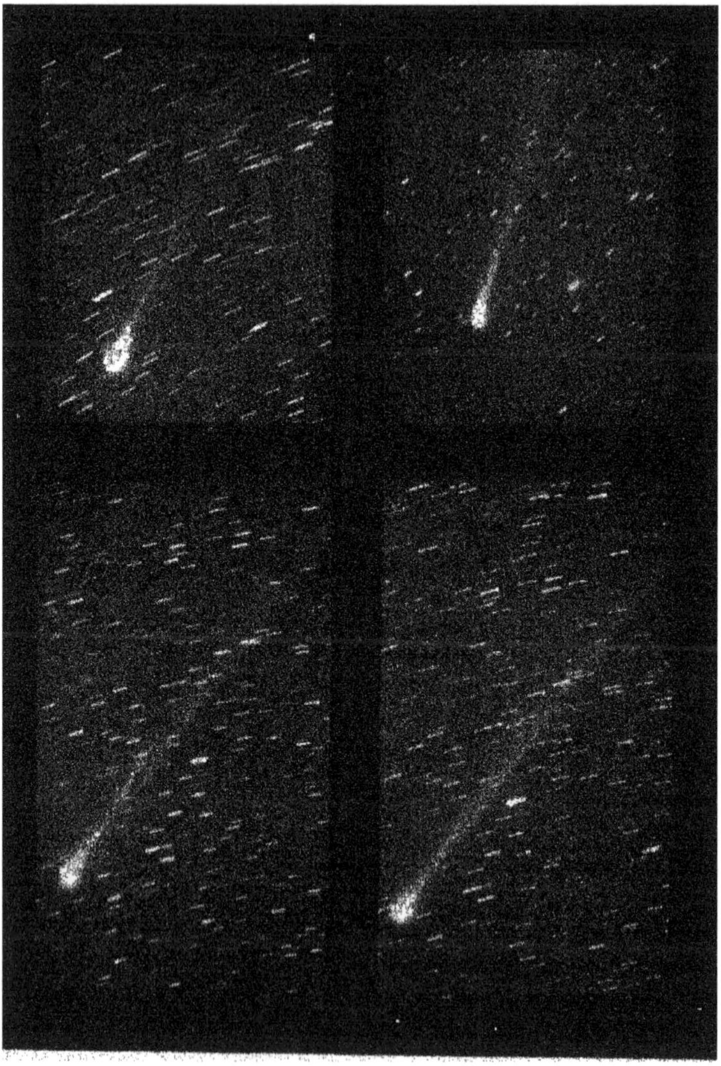

July 24, 17h 50m—20h 29m July 24, 14h 57m—17h 34m

COMET c 1903 (BORRELLY)

Photographed by E E Barnard at Yerkes Observatory

As a final hint to observers, if (4) is not approximately fulfilled it is useless to attempt the solution for the heights. For either it is a question of two different meteors or else the errors of observation are abnormally large. In general the end point is much more accurately observed than the beginning point, as it is obvious that in this case one has his attention already fixed, while in the first case no one knows where or when the meteor will start. Hence it always is probable that it has travelled at least a short distance before it catches the attention of the observer. The main trouble with duplicate observations from two stations is that one observer sees the meteor slightly earlier than the other, hence his beginning point for it does not correspond to the same absolute point in space as does the beginning point for the other. This difficulty, rather than inability to plot the path accurately as actually seen, is the chief reason why (4) is not more exacty fulfilled, at least in the average case which comes up for discussion. There are many devices, which can be employed by the skilled computer, that frequently will permit admissible corrections to be made to the observations from one of the stations. In such cases better results sometimes may be obtained than by using the data as actually furnished.

For brilliant meteors or fireballs, some of which are seen in daylight when there are no stars to serve as reference points, and which may have very many observers in as many different places, the problem is much more complicated. The method developed by von Niessl and used with such great skill and success by him, and more recently successfully applied to such cases by Hoffmeister, is perhaps the best available. Numerous examples by the former may be found in *Sitzungsberichte der Math.- Nat. Akademie der Wissenschaften* of Vienna. An excellent one is given in vol. 121, 1912. Similar examples may also be found in numerous volumes of the *Astronomische Nachrichten*. Calculations of real heights and orbits by Hoffmeister have been appearing for the past ten years or more in a number of German publications.

It is urged that whoever sees a daylight fireball attempt to fix the path with reference to some house, pole, tree or other fixed object and carefully mark the spot on which he was standing. Then by means of a theodolite, or other simple device for measuring angles, the altitudes and azimuths of such points may be determined at leisure. Observations secured in this manner would be most useful in computing the path of such a body in our atmosphere, and in favorable cases also would give us means of computing its orbit in space. In this country the American Meteor Society acts as a repository for such records and attempts the computation for those cases in which enough observations are reported to make computation possible. In England the British Astronomical Association publishes yearly reports on this very subject, while on the Continent there are a number of observatories which are ready and willing to discuss such data reported to them.

Computation of Real Height; Example.

Station 1 McCormick Observatory, Va. $\lambda_1 = 78^\circ\ 31'\ 18''$; $\phi_1 = +38^\circ\ 2'\ 01''$,
Station 2 Washington, D C. $\lambda_2 = 77\ \ 2\ \ 34$, $\varphi_2 = +38\ \ 56\ \ 04$,
The constants were calculated for midnight at S_2 where $\theta_1 = 21^h\ 16^m 30^s$, $\theta_2 = 21\ 22\ 25$ } 1921 Aug 10, $\phi_1' = +37\ 50\ 47$, $\phi_2' = +38\ 44\ 45$,

With these constants equations (1) . (6) were formed as follows:

(1) $x_1 = +0.59636$, (4) $x_2 = +0.60194$, (7) $x_2 - x_1 = +0\ 00558$,
(2) $y_1 = -0.51613$, (5) $y_2 = -0.49404$, (8) $y_2 - y_1 = +0\ 02177$,
(3) $z_1 = +0.61279$, (6) $z_2 = +0\ 62504$, (9) $z_2 - z_1 = +0.01225$,

Solving (7), (8) and (9) for K, D_2 and A_2 we derive the constants.

$\log K = 8.40816$, $D_2 = 28^\circ\ 36'$ and $A_2 = 75^\circ\ 37.5'$ $\therefore\ \theta_1 - A_2 = 243^\circ\ 30'$

The end point of the meteor will be computed. Its observed
coordinates were from S_1, $\alpha_1 = 334^\circ\ 36'$, $\delta_1 = +23^\circ\ 06'$; $\theta_1 = 20^h\ 33^m\ 56^s$,
coordinates were from S_2, $\alpha_2 = 278\ \ 48$; $\delta_2 = -11\ \ 18$; $\theta_2 = 20\ \ 39\ \ 51$.

For this meteor $\therefore\ \theta_1 - A_2 = 243^\circ\ 30'$ also and $\therefore\ A_2 = 64^\circ\ 59'$.

We now form (19) in which $\alpha_2 - A_2 = 213^\circ\ 49'$,
$\alpha_2 - \alpha_1 = -55\ \ 48$,
$\alpha_1 - A_2 = 269\ \ 37$,

The terms of (19) come out respectively $-.396\ +.237\ +.200 = 0$ or $+.396 = +.437$, which is a fairly good check. Therefore the positions doubtless refer to the same meteor.

$\log K$	8 40816	$\log K$	8.40816
$\cos D_2$	9 94351	$\cos D_2$	9.94351
$\sin(\alpha_2 - A_2)$	9.74549n	$\sin(\alpha_1 - A_2)$	9.99999n
Σ_1	8.09716n	Σ_1	8.35166n
$\cos \delta_1$	9.96370	$\cos \delta_2$	9.99150
$\sin(\alpha_2 - \alpha_1)$	9.91755n	$\sin(\alpha_2 - \alpha_1)$	9.91755n
Σ_2	9.88125n	Σ_2	9.90905n
$\rho_1 = \Sigma_1 \div \Sigma_2$ (29)	8 21591	$\rho_2 = \Sigma_1 \div \Sigma_2$ (30)	8.44261

The values of ρ_1 and ρ_2, from (29) and (30) having been derived we are next to substitute these in (36), (37) and (38).

log R_1	9 99946	log R_2	9 99943
cos φ_1	9.89745	cos φ_2	9 89206
cos θ_1	9 79399	cos θ_2	9 80777
log Σ_1	9 69090	log Σ_1	9 69926
Σ_1	+0 49081	Σ_1	+0.50033
log ρ_1	8 21591	log ρ_2	8 44261
cos δ_1	9 96370	cos δ_2	9 99150
cos α_1	9 95585	cos α_2	9 18465
log Σ_2	8 13546	log Σ_2	7 61876
Σ_2	+0 01366	Σ_2	0 00416
$\Sigma_1+\Sigma_2$ = (36)	+0.50447	$\Sigma_1+\Sigma_2$ = (36)	0.50449
log $(\Sigma_1+\Sigma_2)$	9 70284	log $(\Sigma_1+\Sigma_2)$	9.70285
log R_1	9 99946	log R_2	9 99943
cos φ_1	9 89745	cos φ_2	9.89206
sin θ_1	9 89364n	sin θ_1	9 88447n
log Σ_3	9 79055n	log Σ_3	9 77596n
Σ_3	−0 61738	Σ_3	−0 59699
log ρ_1	8.21591	log ρ_2	8 44261
cos δ_1	9 96370	cos δ_2	9 99150
sin α_1	9 63239n	sin α_2	9 99486n
log Σ_4	7 81200n	log Σ_4	8 42897n
Σ_4	−0 00649	Σ_4	−0.02685
$\Sigma_3+\Sigma_4$ = (37)	−0 62387	$\Sigma_3+\Sigma_4$ = (37)	−0 62384
log $(\Sigma_3+\Sigma_4)$	9 79509n	log $(\Sigma_3+\Sigma_4)$	9.79507n
log R_1	9 99946	log R_2	9.99943
sin φ_1	9 78785	sin φ_2	9.79648
log Σ_5	9 78731	log Σ_5	9 79591
Σ_5	+0 61279	Σ_5	+0 62504
log ρ_1	8 21591	log ρ_2	8 44261
sin δ_1	9 59366	sin δ_2	9 29214n
log Σ_6	7.80957	log Σ_6	7.73475n
Σ_6	+0 00645	Σ_6	−0.00543
$\Sigma_5+\Sigma_6$ = (38)	+0 61924	$\Sigma_5+\Sigma_6$ = (38)	+0.61961
log $(\Sigma_5+\Sigma_6)$	9 79186	log $(\Sigma_5+\Sigma_6)$	9 79212
(37) ÷ (36)	0 09225	(37) ÷ (36)	0 09222
α	308° 57 6′	α	308° 57 7′
cos α	9 79850	cos α	9.79852
(36)	9 70284	(36)	9 70285
R cos δ	9 90434	R cos δ	9 90433
(38)	9 79186	(38)	9 79212
tan δ	9 88752	tan δ	9 88779
δ	37° 39.8′	δ	37° 40.8′
sin δ	9 78605	sin δ	9 78621

log R	0.00581	log R	0 00591
log $(R-R_1)$	8.1679	log $(R-R_1)$	8.1767
log R_{eq}	3 8046	log R_{eq}	3.8046
log h	1.9725	log h	1 9813
h (km.)	93.86	h (km)	95 79
Adopted h	94 8		

It scarcely is necessary to point out the excellent check on one's results given by this method of double computation. The results for the beginning point of this same meteor came out for S_1, 124.0 km., for S_2 131.5 km., mean 127.8 km., which is in good agreement also. The computations are not given here as there is no difference whatever, only the coordinates of the beginning point, instead of the end point, are to be used for (α_1, δ_1) and (α_2, δ_2). The meteor was observed by Mr. Donald Brooks at Washington and by the writer at the McCormick Observatory.

CHAPTER XV

Computation of Orbits of Meteors

For the short time during which a meteor is within the (appreciable) sphere of attraction of the earth its motion may be considered as rectilinear and uniform, no matter what sort of a conic it may be describing about the sun. By the earth's attraction its original velocity v will be increased and it will have its orbit changed into a hyperbola, with the earth's center as focus In figure 8 let O be the earth's center,[1] QCD the original direction of motion of a meteor, which due to the earth's attraction will actually be turned into the hyperbola SMF, cutting the earth's surface at M, and which has QCD for an asymptote. The actual direction of motion of the meteor at M will of course be along the tangent VMP. Since O is the focus of the new orbit and QCD is the original direction of motion and an asymptote to the new orbit, the plane of this orbit is sufficiently defined by this point and this straight line, and, further, all the other lines in the figure must of necessity lie in the same plane, which is a vertical plane for the point M, having Z for its zenith Now let $\angle VMZ = z$, the apparent zenith distance; $\angle NMZ = \zeta$ the true zenith distance; and $\zeta - z = \phi = \angle VMN$. At once we see that unless the meteor actually falls in the vertical line ZM at a place M the effect of the earth's attraction is always to make it appear nearer the zenith since z is always smaller than ζ. This is the so-called zenith attraction. Now draw the other asymptote DD′ and prolong OD to L, making $OD = DL$. Let $OM = \rho$, $ML = \rho'$, where $\rho' - \rho = 2a$, and complete figure.

Analytically, at the distance of one astronomical unit the earth's attraction is k^2M, where k^2 is the Gaussian constant, and M the earth's mass in terms of that of the sun. Therefore at M, which is at a distance ρ from the earth's center, the attraction g amounts to $\frac{k^2M}{\rho^2}$, ρ again being expressed in astronomical units. Therefore

$$k^2M = g\rho \tag{1}$$

[1] *Sternschnuppen, Note* 4

For motions in a hyperbola, at a point with radius vector r, we have $v^2 = k^2M\left(\frac{2}{r} + \frac{1}{a}\right)$ or substituting

(2) $$v^2 = g\rho^2\left(\frac{2}{r} + \frac{1}{a}\right)$$

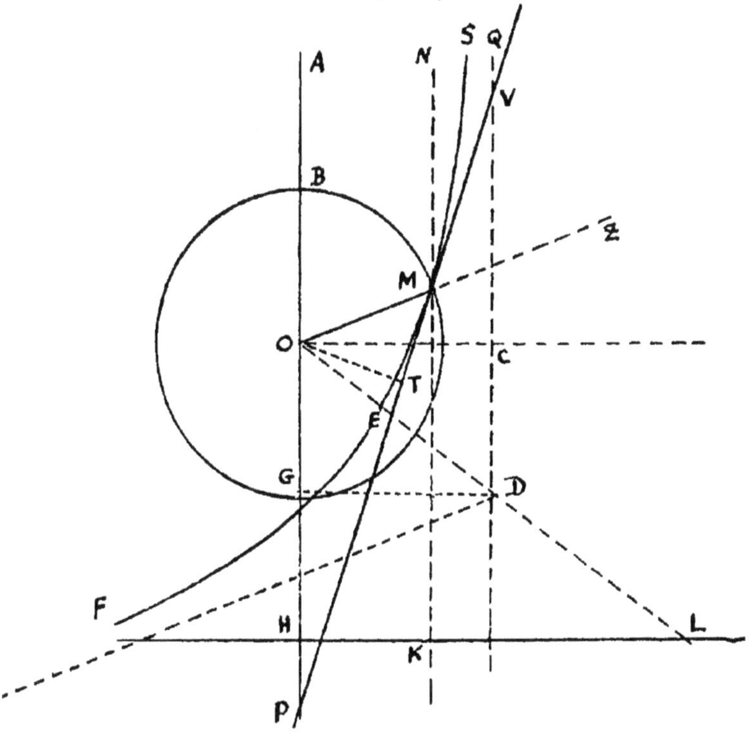

Fig. 8

If u represents the velocity from infinity, r = ∞ and this expression becomes

(3) $$u^2 = \frac{g\rho^2}{a}$$

Again if w represents the velocity at a distance ρ, we have

(4) $$w^2 = g\rho^2\left(\frac{2}{\rho} + \frac{1}{a}\right)$$

Substituting we get

$$w^2 = u^2 + 2g\rho \tag{5}$$

From observations upon the earth we have found that $\rho = 6.37 \times 10^6$ meters, $g = 9.8$ meters per second, $\therefore\ 2g\rho = 124852000\left(\frac{\text{met.}}{\text{sec.}}\right)^2$. It is therefore clear that if w, which is the apparent velocity, can be accurately observed, u can at once be found. Or, vice versa, if either can be found from other data or theoretical considerations, the other can be computed. When u has been computed the semi-major axis of the hyperbola is at once found from

$$a = g\left(\frac{\rho}{u}\right)^2 \tag{3}$$

Again we have the general proposition that for any point P at a given distance x from the sun (or any attracting body in terms of the sun's mass) when the direction and amount of its motion is known, then its orbit can be obtained. Or

$$n = \frac{k\sqrt{M + m}\,\sqrt{p}}{v} \tag{6}$$

where n is the perpendicular to the tangent at point P, p is the parameter of the orbit, and v the velocity in the orbit. In our case $m = 0$, as the meteor's mass is vanishingly small, $\therefore\ k\sqrt{M}\,\sqrt{p} = OT.w = (\rho \sin z)w$.

But $p = a(e^2 - 1)$ and $g = -\frac{k^2M}{\rho^2}\ \therefore$

$$a(e^2 - 1) = \frac{\rho^2 \sin^2 z \cdot w^2}{k^2 M} = \frac{\sin^2 z \cdot w^2}{g} \tag{7}$$

From the properties of the hyperbola we also know $e = \sec\psi$. Now substituting (3) in (7) we have

$$\tan\psi = \frac{u\,w}{g\,\rho}\sin z \tag{8}$$

Let $OC = s$, i.e., the conjugate axis. Then noting that $CD = a$, because $CD = DR = OD\cos\psi = ae\cos\psi = a$,

$$s = a\tan\psi = \rho\frac{w}{u}\sin z \tag{9}$$

Also $LM = \rho + 2a$. Since the tangent at M bisects $\angle LMO$, we have $\angle LMT = \angle TMO = \angle VMZ = z$, $\angle TMK = \angle VMN = \phi$.

And $\angle BOM = \angle NMZ = \zeta \therefore \angle LMK = z - \phi = \theta$, $\angle OMK = z + \phi$, $MK = IK + MI = 2a + \rho \cos \zeta$. But

(10) $$MK = ML \cos\theta = (\rho + 2a)\cos(z - \phi) = 2a + \rho \cos \zeta$$

As $w^2 = u^2 - 2g\rho$ and $a = g\left(\frac{\rho}{u}\right)^2$, equation (10) becomes $w^2 \cos\theta = 2g\rho + u^2 \cos \zeta$. Also $\angle OML = \angle OMK + \angle KML = \zeta + \theta$, and $\angle OML = 2\,\angle OMT = 2z \therefore 2z = \zeta + \theta$ or $\theta = 2z - \zeta$ $\therefore w^2 \cos(2z - \zeta) = 2g\rho + u^2 \cos \zeta$. Adding w to each side this becomes $w^2 [1 - \cos(2z - \zeta)] = u^2(1 - \cos \zeta)$ or

(11) $$w^2 \sin(z - \tfrac{1}{2}\zeta) = u^2 \sin \tfrac{1}{2}\zeta$$

From this relation z can at once be computed when ζ is known. However the inverse problem is the one which is met with in practice. As $\zeta = z + \phi \therefore w^2 \sin \tfrac{1}{2}(z - \phi) = u^2 \sin \tfrac{1}{2}(z + \phi)$ or

(12) $$\tan \tfrac{1}{2}\phi = \frac{w - u}{w + u} \tan \tfrac{1}{2} z$$

It is therefore obvious that to find the zenith attraction ϕ it is necessary to know both the velocities u and w, as well as the observed or apparent zenith distance. It is further seen that ϕ is a maximum when $z = 90°$ and that $\phi = 0$, for $z = 0°$, i e., in the zenith. For parabolic velocity, $u = 42$ km./sec., $\phi = 17° +$ at $z = 90°$, as a maximum Fortunately for most positions, particularly within 90° of the apex and if the radiant is not too near the horizon, the value of ϕ is small The next table, copied from *Sternschnuppen,* will give the values of ϕ for different values of z and different distances from the apex. This correction ϕ is the most important to be applied to the observed position of the radiant.

Diurnal aberration. The regular formulae for diurnal aberration are:

$$\begin{cases} \alpha' - \alpha = 0.32'' \cos\phi \cos t \sec\delta \\ \delta' - \delta = 0.32'' \cos\phi \sin t \sin\delta \end{cases}$$

where $t = \theta - \alpha$ and ϕ is the latitude of the place of observation.[2] These formulae serve our purposes if only we put in the proper constant instead of 0.32″ which shall correspond to the case of meteors. This can be done with all sufficient accuracy by substituting

[2] The θ and ϕ used in the discussion of diurnal aberration are the conventional symbols for sidereal time and geographical latitude respectively and are not to be mistaken for the angles θ and ϕ as defined in the previous pages.

$\left(\frac{2\pi\rho}{86400}\right)\left(\frac{57.3^\circ}{w}\right)$ where ρ is the radius of the earth in kilometers and w the observed velocity of the meteor expressed in kilometers per second. Expressed in angular measure this becomes $\left(\frac{26\ 57}{w}\right)^\circ$ which for parabolic velocity is about 0.6°+. The necessary formulae are therefore

$$(13)\quad \begin{cases} \alpha' - \alpha = \left(\frac{26.57}{w}\right)^\circ \cos\phi \cos t \sec\delta \\ \delta' - \delta = \left(\frac{26\ 57}{w}\right)^\circ \cos\phi \sin t \sin\delta \end{cases} \quad \text{or} \quad \begin{cases} \Delta\alpha = -\left(\frac{26\ 57}{w}\right)^\circ \cos\phi \cos t \sec\delta \\ \Delta\delta = -\left(\frac{26\ 57}{w}\right)^\circ \cos\phi \sin t \sin\delta \end{cases}$$

It is obvious that the effect is a maximum at the earth's equator where $\phi = 0^\circ$, and a minimum at $\phi = 90^\circ$, or the poles. It is further seen that the maximum value of $\delta' - \delta$ is only the bracket term itself. For $\alpha' - \alpha$, where δ is very great, sec δ may attain large values, but not very many radiants will ever be found close to the celestial poles, where the correction would be considerable.

APPARENT ELONGATION FROM APEX	$\log \frac{u}{w}$	APPARENT ZENITH—DISTANCE OF THE RADIANT									
		Z=0°	Z=10°	Z=20°	Z=30°	Z=40°	Z=50°	Z=60°	Z=70°	Z=80°	Z=90°
0°	9 99463	0° 0′	0° 4′	0° 7′	0° 11′	0° 15′	0° 20′	0° 24′	0° 30′	0° 36′	0° 42′
12	9 99447	0 0	0 4	0 8	0 12	0 16	0 20	0 25	0 31	0 37	0 44
24	9 99394	0 0	0 4	0 8	0 14	0 17	0 22	0 28	0 34	0 40	0 48
36	9 99291	0 0	0 5	0 10	0 15	0 20	0 20	0 32	0 47	0 47	0 56
48	9 99116	0 0	0 6	0 12	0 19	0 25	0 33	0 40	0 49	0 59	1 10
60	9 98824	0 0	0 8	0 16	0 25	0 34	0 43	0 53	1 05	1 18	1 33
72	9 98344	0 0	0 11	0 23	0 35	0 47	1 01	1 15	1 31	1 49	2 11
84	9 97573	0 0	0 17	0 34	0 51	0 10	1 30	1 51	2 14	2 41	3 12
96	9 96415	0 0	0 25	0 50	1 16	1 23	2 12	2 44	3 19	3 58	4 43
108	9 94852	0 0	0 36	1 12	1 49	2 28	3 10	3 55	4 45	5 41	6 46
120	9 92975	0 0	0 49	1 38	2 29	3 36	4 19	5 20	6 29	7 45	9 14
132	9 91033	0 0	1 02	2 05	3 09	4 17	5 30	6 48	8 15	9 54	11 44
144	9 89268	0 0	1 14	2 29	3 46	5 07	6 34	8 07	9 53	11 47	14 01
156	9 87855	0 0	1 24	2 48	4 16	5 47	7 26	9 10	11 07	13 18	15 49
168	9 86957	0 0	1 30	3 01	4 34	6 13	7 58	9 52	11 55	14 15	16 57
180	9 86650	0 0	1 32	3 05	4 41	6 21	8 08	10 04	12 11	14 35	17 20

Note: For streams in parabolic orbits the table can be entered with either the apparent elongation from the apex or $\log \frac{u}{w}$: for elliptical or hyperbolic streams only with $\log \frac{u}{w}$.

As the effect of diurnal aberration is to increase the apparent or observed right ascension the correction is to be subtracted from the observed right ascension, unless cos t becomes negative. $\Delta\delta$ will become positive when t is in the third or fourth quadrant, or if δ has the minus sign.

When the observed radiant has its position corrected for both zenith attraction and diurnal aberration, we then obtain the corrected apparent radiant which should, in cases where the greatest accuracy is necessary, be used in the computation of orbits. Fortunately in very many cases, particularly for radiants secured after midnight, both corrections are smaller than the probable errors of observation and hence can be most frequently neglected.

To show the effects of these corrections in two very typical and important instances we quote from an article of much importance by Kleiber[3] in which were calculated the effects of the attraction, rotation and orbital motion of the earth upon meteor streams. The results are of such fundamental illustrative interest that we give part of them in a condensed form. For his purposes he chose the Bielids and the Perseids, both so much studied as we have seen in previous pages. He calculated the zenith correction by Schiaparelli's formula $\Delta z = c \tan \frac{1}{2} z$ and the diurnal aberration by the formulae of Lehmann Filhès. He calls these corrections Δ_1 and Δ_2 respectively, that due to the motion of the earth Δ_3. We abridge his tables, which are calculated for the latitude of Greenwich, very considerably.

Perseids: $\alpha_0 = 43.58°$; $\delta = +57\ 08°$; for August 9 12^h

Date		$\Delta\alpha_3$	$\Delta\alpha_1+\Delta\alpha_2$	$\Sigma\Delta\alpha$	$\Delta\delta_3$	$\Delta\delta_1+\Delta\delta_2$	$\Sigma\Delta\delta$	Corrected Radiant α	δ
Aug 8	0^h	−2 01°	+0 71°	−1 30°	−0 42°	+0 03°	−0 39°	42 28°	+56 69°
	12	−1 34	−0 71	−2 05	−0 28	+0 49	+0 21	41 53	57 29
Aug. 9	0	−0 67	+0 71	+0 04	−0 14	+0 03	−0 11	43 62	57 97
	12	0 00	−0 71	−0 71	0 00	+0 49	+0 49	42 87	57 57
Aug. 10	0	+0 67	+0 71	+1 38	+0 14	+0 03	+0 17	44 96	57 25
	12	+1 34	−0 71	+0 63	+0 28	+0 49	+0 77	44 21	57 85
Aug 11	0	+2 01	+0 71	+2 72	+0 42	+0 03	+0 45	46 30	57 53

[3] *Monthly Not., R.A.S.*, **52**, 341, 1892.

Bielids: $\alpha_0 = 23.27°$; $\delta_0 = +\ 43.12°$; for November 27 12h

Date		$\Delta\alpha_2$	$\Delta\alpha_1+\Delta\alpha_2$	$\Sigma\Delta\alpha$	$\Delta\delta_2$	$\Delta\delta_1+\Delta\delta_2$	$\Sigma\Delta\delta$	Corrected Radiant α	δ
Nov. 26	0h	−1 62°	−4 69°	−6 31°	−0 87°	+8 77°	+7 90°	16 96°	51 08°
	12	−1 08	+2 73	+1 65	−0 58	+1 20	+0.62	24 92	43 78
Nov. 27	0	−0 54	−4 69	−5 23	−0 29	+8 77	+8 48	18 04	51 60
	12	0 00	+2 73	+2 73	0 00	+1 20	+1 20	26 00	44 32
Nov. 28	0	+0 54	−4 69	−4 15	+0 29	+8 77	+9 06	19 12	52 18
	12	1 08	+2 73	+3 81	+0 58	+1 20	+1 78	27 08	44 90
	15	1 22	+6 65	+7 87	+0 65	+3 64	+4 29	31 14	47 41
	18	+1 35	+6 75	+8 10	+0 72	+7 55	+8 27	31 37	51 39
	21	+1 48	+1 45	+2 93	+0 80	+10 80	+11 60	26 20	54 72

Comment on these results is almost superfluous, nevertheless it is desired to reiterate the statements contained in Chapter IX as to the difficulty of determining real (instead of spurious) radiants anywhere near the anti-apex, and the great need of using observations included within a few hours only of a given night for the determination of radiants.

Having now shown how the observed radiant can be freed from the harmful effects of the attraction and rotation of the earth, we pass to the methods used for the computation of the true orbit of a meteor in space. We will assume that the position of the radiant to be discussed has already been corrected for these two effects just mentioned.

In figure 9[4] let the vector OB represent the velocity of the earth = V, DB the real velocity of the meteor = v, and therefore C′B the apparent velocity of the meteor = w. We then have C the true radiant point, C′ the apparent radiant point, and A the meteoric apex or the point toward which the earth is moving at the moment. Let $\angle ABC = \chi$ and $\angle ABC' = \sigma \therefore \angle C'BC = \chi - \sigma$.

Now in figure 10, which represents the celestial sphere, let γ AHK represent the plane of the earth's orbit or the ecliptic, and P its pole. With the earth at center B draw the radii BA, BC′, and BC when A represents the position of the apex, C′ that of the apparent radiant and C that of the true radiant. The angles χ, σ, and $\chi - \sigma$ can now be referred to this figure, the meaning remaining unchanged. It is obvious that A, C′ and C must lie upon the arc of a great circle,

[4] The development of the formulae for computation of meteor orbits is mostly taken from *Die Bahnbestimmung Von Meteorbahnen* by Lehmann-Filhès.

because BC′ must lie in the plane defined by the intersecting lines BA and BC, and this plane cuts the celestial sphere in arc AC′C.

If the earth's orbit were a perfect circle then the position of A would always be $L = \odot - 90°$. But due to the fact that the orbit is actually an ellipse, the term 90° may vary by ±1° In careful researches it is therefore necessary to derive an accurate value of L.

If R and L′ be the radius vector of the earth's orbit and its longitude then $L' = 180° + \odot$ in all cases. The velocity components can then be expressed as

$$(14)\quad \begin{cases} V\cos L = \frac{d}{dt}(R\cos L') = -\frac{d}{dt}(R\cos\odot) \\ V\sin L = \frac{d}{dt}(R\sin L') = -\frac{d}{dt}(R\sin\odot) \end{cases}$$

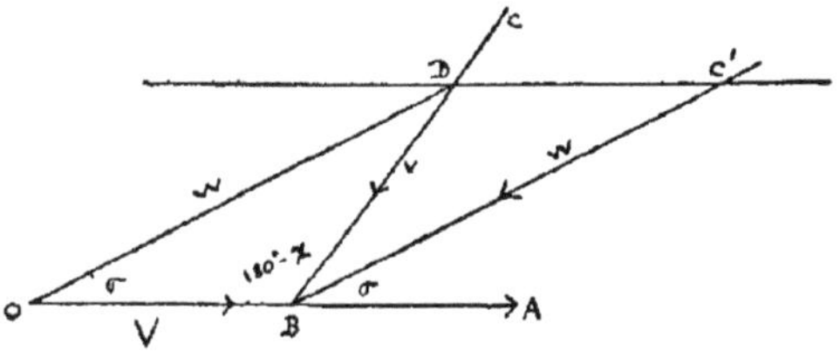

FIG. 9

Differentiate the last term in each and multiply the upper by $\cos\odot$, the lower by $\sin\odot$ and add. We then at once obtain

$$(15)\quad \begin{cases} V(\cos L\cos\odot + \sin L\sin\odot) = -(\cos^2\odot + \sin^2\odot)\frac{dR}{dt} \\ V\cos(L - \odot) = -\frac{dR}{dt} \end{cases}$$

Now multiply by $\sin\odot$ and $-\cos\odot$ respectively. Reducing we find

$$V\sin(L - \odot) = -R\frac{d\odot}{dt}$$

$$\therefore \tan(L - \odot) = R\frac{d\odot}{dR}$$

Also we know that

$$R = \frac{a(1 - e^2)}{1 + e\cos(L' - \omega)} = \frac{1 - e^2}{1 - e\cos(\odot - \omega)}$$

where a = 1, e is the eccentricity, and ω the longitude of perihelion of the earth's orbit. As e = 0.0168 only, the term e^2 may safely be neglected.

$$\therefore R = 1 + e \cos (\odot - \omega)$$

$$\frac{dR}{d\odot} = - e \sin (\odot - \omega)$$

$$\therefore \tan (L - \odot) = - \frac{R}{e \sin (\odot - \omega)}$$

$$\tan (90^\circ + L - \odot) = \frac{e \sin (\odot - \omega)}{R}$$

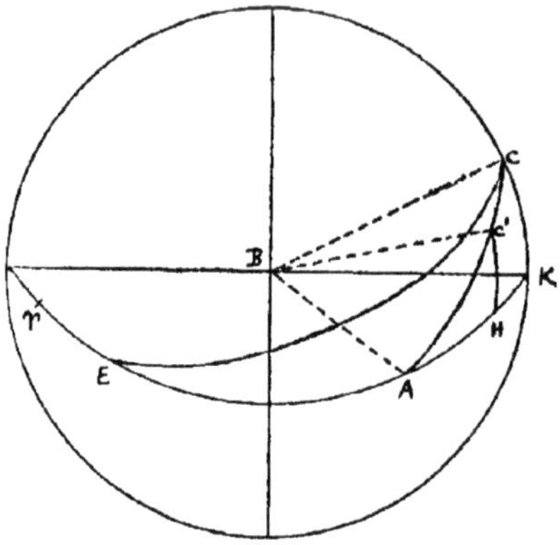

FIG. 10

or since R = 1 practically, and turning into arc we get,

$$(16) \qquad L = \odot - 90^\circ = \frac{e}{\sin 1'} \sin (\odot - \omega)$$

In this $\omega = 101^\circ\ 13'.2 - 1.028'\ (t - 1900)$ and $\frac{e}{\sin 1'} = 57'.6$.

Having found an accurate value for L, we can proceed to the actual derivation of the elements of the meteor orbit. In figure 10 let γ be the first of Aries, L' the longitude of the earth = 180° + $\odot$. Draw arcs CK and C'H perpendicular to the ecliptic. Then we have the coordinates of C and C' as follows: $l = \gamma K$, $b = KC$, $\lambda = \gamma H$, and $\beta = HC'$. Also let $\gamma A = L$ and $\angle KAC = \gamma$. Then in the two right spherical triangles AHC′ and AKC we have from elementary trigonometry:

$$(17)\begin{cases}\cos\sigma = \cos\beta\cos(\lambda - L)\\ \sin\gamma\sin\sigma = \sin\beta\\ \cos\gamma\sin\sigma = \cos\beta\sin(\lambda - L)\end{cases} \qquad (18)\begin{cases}\cos\chi = \cos b\cos(l - L)\\ \sin\gamma\sin\chi = \sin b\\ \cos\gamma\sin\chi = \cos b\sin(l - L)\end{cases}$$

We shall always keep σ so that $180° > \sigma > 0°$, and as $\sin\gamma\sin\sigma = \sin\beta$, and as $\sin\sigma$ is always positive, therefore $\sin\gamma$ must be negative when $\sin\beta$ is negative or in general $360° > \gamma > 0°$, so that $\sin\gamma$ may take both signs as required. Also from figure 9 we have the following relations, since OBC is a plane triangle

$$(19)\quad\begin{cases}v^2 = w^2 - V^2 - 2wV\cos\sigma\\ v\sin(\chi - \sigma) = V\sin\sigma\\ \text{or } \dfrac{V}{v} = \dfrac{\sin(\chi - \sigma)}{\sin\sigma}\end{cases}$$

Finally for the earth we have $V^2 = k^2\left(\frac{2}{R} - 1\right)$ for the meteor $v^2 = k^2\left(\frac{2}{R} - \frac{1}{a}\right)$. Equating and taking the square root we find $\frac{V}{v} = \sqrt{\frac{\frac{2}{R} - 1}{\frac{2}{R} - a}}$, an expression which we have already found is equal to $\frac{\sin(\chi - \sigma)}{\sin\sigma}$. In other words if we know v from either observation or theoretical considerations, the problem can be solved. Obviously for the parabola, where $a = \infty$, the expression takes the very simple form

$$(20)\qquad \frac{V}{v} = \sqrt{1 - \frac{R}{2}}$$

Since the value for R is given in the American Ephemeris to any desired degree of accuracy, the corresponding value of V can at once be found. For the convenience of those without proper tables we give the following values. Log k = 8.2355814 or k = 0.01720210. The units used are a, the astronomical unit, which equals 1; M, the mass of the sun, also taken as 1; m, the earth's mass, in terms of the sun's mass; T = 365.25638 days. The equation from which k is derived is $k = \frac{2\,a^{3/2}\,\pi}{T\sqrt{M+m}}$.

Let us now call the distance from the true radiant C to the earth $180^\circ - \eta$. Then in triangle EAC we also have $AC = \chi$, $EA = L - (180^\circ + \odot)$ and $\angle\gamma\ EC = \iota$, which is the inclination of the meteor's orbit to the ecliptic. The obtuse angle is chosen because E is at the descending node of the meteor's orbit in accord with the way the figure has been drawn.

To make the relations clearer figure 11 is also given, the upper part corresponding to the case now considered for $\Omega = \odot$. Then we obtain

$$
(21)\quad \begin{cases} \sin\iota\sin\eta = \sin\gamma\sin\chi \\ -\cos\iota\sin\eta = +\cos\chi\sin(\odot - L) - \cos\gamma\sin\chi\cos(\odot - L) \\ \cos\eta = \cos\chi\cos(\odot - L) + \sin\chi\sin(\odot - L)\cos\gamma \end{cases}
$$

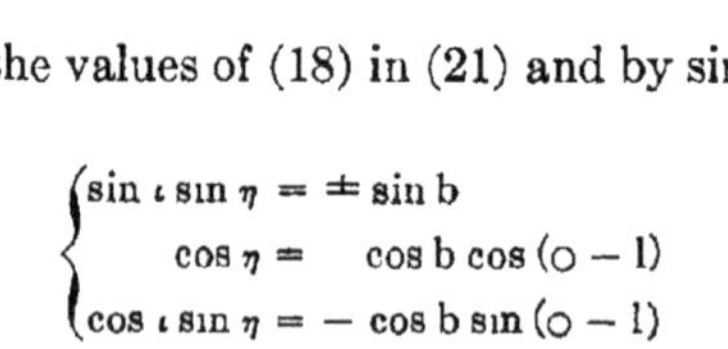

Fig 11

By substituting the values of (18) in (21) and by simple reduction we derive

$$
(22)\quad \begin{cases} \sin\iota\sin\eta = \pm\sin b \\ \cos\eta = \cos b\cos(\odot - l) \\ \cos\iota\sin\eta = -\cos b\sin(\odot - l) \end{cases}
$$

If the true radiant C is south of the ecliptic, then $\Omega = 180^\circ + \odot$, since the meteor meets the earth in the ascending node. In this case consider the lower part of figure 11. Here C_2 represents the true radiant, but since its latitude is south and therefore negative, we count $C_2A\odot = 360^\circ - \gamma$. Then from either triangle EAC_2 or $C_2A\odot$ we obtain as before

$$
(23)\quad \begin{cases} \cos\eta = \cos\chi\cos(\odot - L) + \sin\chi\sin(\odot - L)\cos\gamma = \cos b\cos(\odot - l) \\ \sin\iota\sin\eta = \sin\chi\sin\gamma = -\sin b \\ -\cos\iota\sin\eta = \cos\chi\sin(\odot - L) - \sin\chi\cos(\odot - L)\cos\gamma = \cos b\sin(\odot - l) \end{cases}
$$

It will be seen that the last terms of (23) are exactly the same as in (22) except for the $-$ sign before sin b. Hence in solving (22) or (23) in the first case for $\Omega = \text{o}$, the $+$ sign is to be used before sin b; in the second case for $\Omega = \text{o} + 180$, the $-$ sign is to be used before sin b. It is quite obvious that in each case the apparent radiant and the true radiant must both lie on the same side of the ecliptic.

So far the equations developed are equally good for every kind of conic section and have given us the means of computing ι and Ω, but as in practice we have three different kinds of orbits to deal with, from here on they must be treated separately. As the parabola is the simplest case it will first be considered.

If the meteor has already passed perihelion, let it meet the earth in E. In figure 12 draw o E, also let EQ be parallel to π o, the

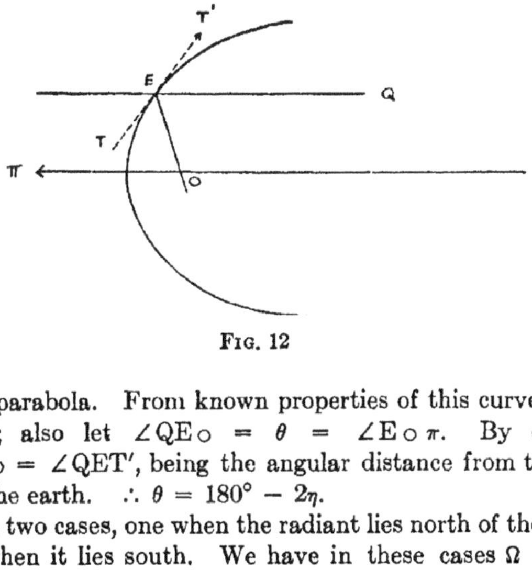

FIG. 12

axis of the parabola. From known properties of this curve $\angle$T'EQ $= \angle$TE o; also let $\angle$QE o $= \theta = \angle$E o π. By definition $\eta = \angle$TE o $= \angle$QET', being the angular distance from the sun as seen from the earth. $\therefore \theta = 180° - 2\eta$.

There are two cases, one when the radiant lies north of the ecliptic, the other when it lies south. We have in these cases $\Omega = \text{o}$ and $\Omega = 180° + \text{o}$ respectively.

If the perihelion has not been passed, θ is negative, then

$$\angle \pi \text{o E} = \angle \text{o EQ} = -\theta; \ \angle \text{o ET} = \eta;$$

also $\angle$ o ET' $= \angle$QET $= 180° - \eta$. $\theta + 2\,(180° - \eta) = 180°$; then $\theta = 180° - 2\,\eta$ which is the same equation as derived for the other case. If the true anomaly of the ascending node be denoted by θ_Ω then π is in general either $\Omega - \theta_\Omega$ or $\Omega + 360° - \theta_\Omega$, both of which are the same.

If the radiant is north of the ecliptic and the earth at the descending node then $\Omega = \odot$, $\theta_\Omega = \theta \pm 180° = 180° - 2\eta \pm 180°$ and $\pi = \odot + 2\eta$, since even multiples of 180° need not be considered.

When the radiant is south of the ecliptic, and the earth in the ascending node,

$$(24) \qquad \begin{cases} \Omega = 180° + \odot, \ \theta_\Omega = \theta = 180° - 2\eta \\ \pi = \odot + 2\eta \end{cases}$$

The cases worked out hold for both direct and retrograde motion. For orbits whose $\iota > 90°$ and in which the motion is therefore retrograde some older writers used exactly the same procedure only the quantity 2η had the minus sign. Therefore in such cases they wrote $\pi = \odot - 2\eta$.

Great care must therefore be exercised in comparing such results with those for which the modern practice is employed. For instance all longitudes of perihelia, copied from Schiaparelli, are expressed in the older way.

To solve for the elements of an elliptical orbit, the analytical method will be followed We have already found the general equation

$$\frac{V}{v} = \sqrt{\frac{\frac{2}{R} = 1}{\frac{2}{R} = \frac{1}{a}}} = \sqrt{\frac{a(2 - R)}{2a - R}}, \text{ also } a^3 = U^2$$

This equation can give us χ and σ, from which η and ι can be calculated as before. But to derive the elements π and $e = \sin \phi$ another procedure is necessary.

With the ecliptic for fundamental plane, sun as origin, + X-axis directed toward the First of Ares, + Y-axis toward Summer Solstice, and + Z-axis toward pole of the ecliptic, then the equations of motion of the meteor at the instant it meets the earth are

$$(25) \qquad \begin{cases} \frac{dx}{dt} = -v \cos b \cos l \\ \frac{dy}{dt} = -v \cos b \sin l \\ \frac{dt}{dt} = -v \sin b \end{cases}$$

The coördinates of the earth at this moment, which are also those of the meteor, are

$$(26)\qquad \begin{cases} x = -\,R\cos\odot \\ y = -\,R\sin\odot \\ z = \quad 0 \end{cases}$$

From the general equations of celestial mechanics, which are used in such cases, we have

$$(27)\qquad \begin{cases} k\sqrt{p}\cos\iota = x\dfrac{dy}{dt} - y\dfrac{dx}{dt} = R\,v\cos b\sin(l - \odot) \\ k\sqrt{p}\sin\Omega\sin\iota = y\dfrac{dz}{dt} - z\dfrac{dy}{dt} = R\ v\ \sin b\ \cos\odot \\ k\sqrt{p}\cos\Omega\sin\iota = x\dfrac{dz}{dt} - z\dfrac{dx}{dt} = R\ v\ \sin b\ \sin\odot \end{cases}$$

the last column derived by the proper substitutions from (25) and (26) in the second column. Also since $\Omega = \odot(b+)$ and

$\Omega = 180 + \odot(b-)$ we obtain

$$(28)\qquad \begin{cases} k\sqrt{p}\cos\iota = -\,R\,v\cos b\sin(\odot - l) \\ \pm\, k\sqrt{p}\sin\iota = +\,R\,v\sin b \end{cases}$$

From the solution of (28) p can be found and an independent value of ι, which must check that already derived in (22). In an ellipse $p = a\cos^2\phi = a\,(1 - e^2)$, where $e = \sin\phi$. When p and a are known, therefore e can be found. However as a check it may also be derived as follows

$$(29)\qquad \cos\phi = \frac{Rv\cos b\sin(l - \odot)}{k\sqrt{a}\cos\iota} = \pm\,\frac{Rv\sin b}{k\sqrt{a}\sin\iota}$$

That $\cos\phi$ must always be positive is obvious, since e is a ratio which is always positive. Also we have the perihelion distance from

$$(30)\qquad q = a\,(1 - e)$$

As for the parabola, $\pi = \Omega - \theta_\Omega$ where θ is the true anomaly. From celestial mechanics we have, for motion in an ellipse,

$$(31)\qquad \begin{cases} e\sin\theta = \dfrac{\sqrt{p}}{k\,R}\left(x\dfrac{dx}{dt} + y\dfrac{dy}{dt} + z\dfrac{dz}{dt}\right) = \dfrac{\sqrt{p}}{k\,R}\,Rv\cos b\cos(l - \odot) \\ e\cos\theta = \dfrac{p}{R} - 1 \end{cases}$$

The last term of the first equation is found by substituting (25) and (26) in the second part. Again from (28)

$$e \sin \theta = \frac{p}{R} \cos \iota \cot (l - \circ) = \frac{p}{R} \cot \eta \therefore$$

$$\tan \theta = \frac{\cot \eta}{1 - R/a \cos^2 \phi} = \frac{\cot \eta}{1 - R/p.}$$

By substitution in $\pi = 180° + \circ - \theta$, π is at once found.

For hyperbolic motion only a few changes are necessary in the equations just derived for the ellipse. In this case as there is no period, the only way a can be found is to observe the velocity v, or to assume it on some theoretical grounds. Since in general $v^2 = k^2 \left(\frac{2}{R} - \frac{1}{a}\right)$, if we let A be the absolute value for a, we have

$$A = \frac{1}{\frac{v^2}{k^2} - \frac{2.}{R}}$$

The parameter p will come from (27)

(32) $$\sqrt{p} = \frac{Rv \cos b \sin (l - \circ)}{k \cos \iota} = \pm \frac{Rv \sin b}{k \sin \iota}$$

and the eccentricity from

(33) $$e = \sqrt{1 + p/A}$$

As before $\pi = 180° + \circ - \theta$.

For the hyperbola we have $\tan \frac{1}{2}v = \tan \frac{1}{2}F \cot \frac{1}{2}\psi$ and $\tan \psi = \sqrt{e^2 - 1}$. Putting θ for v and reducing the value $\cot \frac{1}{2}\psi$ through $\tan \psi = \sqrt{e^2 - 1}$ we get

(34) $$\tan \tfrac{1}{2} F = \tan \tfrac{1}{2} \theta \sqrt{\frac{e - 1}{e + 1}}$$

Finally the time of perihelion passage T can be found from

(35) $$\frac{k}{a^{\frac{3}{2}}} (t - T) = - F + \frac{e}{2} (e_1^F - e_1^{-F}) = - F - e \sinh F$$

in which e_1 represents the natural base of logarithms, the subscript being used to differentiate it from e which denotes the eccentricity.

Notes: The equations and methods employed in this chapter have mostly been taken from the work of Schiaparelli, Lehmann-Filhès, von Niessl, Herz, Bauschinger, Klinkerfues and Moulton. Perhaps the most complete treatment may be found in Bauschinger's *Bahnbestimmung*, in which he indeed largely followed the work of Lehmann-Filhès.

Computation of a parabolic orbit

Radiant α	39° 30′	tan β	9 9220
Radiant δ	+57 48	sin (λ-L)	9 3393
tan δ	0 2008	tan γ	0 5827
sin α	9 8035	γ	75° 21′
tan F	0 3973	cos β	9 8850
F	68° 10′	cos (λ-L)	9 9894
ϵ	23 27	cos n	9 8744
F−ϵ	44 43	n	41° 31′
cos (F−ϵ)	9 8516	sin β	9 8070
tan α	9 9161	sin γ	9 9856
Σ	9 7677	sin n	9 8214
cos F	9 5704	n	41° 31′
tan λ	0 1973	log $\frac{V}{v}$	9 8464
λ	57° 35′		
tan (F−ϵ)	9 9957	sin n	9 8214
sin λ	9 9264	sin (s-n)	9 6678
tan β	9 9221	s−n	27° 44′
β	+39° 53′	s	69 15
Check I		tan s	0 4215
cos δ	9 7266	cos γ	9 4030
sin α	9 8035	tan (l-L)	9 8245
sec F	0 4296	l-L	33° 44′
cos (F−ϵ)	9 8516	l	78 42
cosec λ	0 0736	sin s	9 9709
sec β	0 1151	sin γ	9 9856
log 1	0 0000	sin b	9 9565
Check II		b	64° 47′
cos δ	9 7266	cos s	9 5494
cos α	9 8874	cos (l-L)	9 9199
Σ	9 6140	cos b	9 6295
cos λ	9 7291	b	64° 47′
sec β	0 1151	tan b	0 3270
		l−☉	−55° 45′
Date 1916 Aug. 6 83		tan b	0 3270
☉′	133° 40′	sin (l−☉)	9 9173n
Δ	+47	tan ι	0 4097n
☉′ + Δ = ☉	134° 27′	ι	111° 16′
log R	0 0060	cos b	9 6295

Computation of a parabolic orbit—Continued

ω	101° 29′	cos (l − ☉)	9 7504
☉ − ω	32 58	cos η	9 3799
C	1 7604	η	76° 03′
sin (☉ − ω)	9 7357	sin b	9 9565
log ϵ	1 4961	sin ι	9 9694
ϵ'	+31′	sin η	9 9871
☉ + ϵ'	134° 58′	η	76° 03′
☉ + ϵ' − 90° = L	44 58	$\sin^2 \eta$	9 9742
l′ − L	12 37	log R	0 0060
R	1 0139	log q	9 9802
1 − R/2	0 4930	q	0 9554
log (1 − R/2)	9 6928	2η	−152° 06′
log $\sqrt{1-R/2}$	9 8464	☉	134 27
		π	342 21

CHAPTER XVI

Apparent Distribution of Meteors in Time and Spaces

We have already made a few general statements about the distribution of the numbers of meteors seen as to hour and day of the year. Here a more careful study of the question, along with tables of data will be given.

MONTH	COULVIER-GRAVIER	SCHMIDT	WOLF	DENNING	OLIVIER	HOFF-MEISTER
January	3 6	8 6	5 5	9 2	6 6	7 3
February	3 6	5 6	5 4	7 3	6 2	6 0
March	2 7	6 5	5 2	7 7	6 9	7 7
April .	3 7	6 4	4 6	7 1	8 3	6 9
May	3 8	6 0	4 1	6 0	13 6 *	6 1
June	3 2	6 1	5 4	6 6	11 0	6 0
July... ..	7 0	11 1	9 8	14 3	11 7	12 0
August..	8 5	20 6	12 9	23 7	14 8	9 5
September	6 8	9 8	7 4	13 9	11 6	11 4
October .	9 1	14 2	6 4	15 8	12 2	14 5
November	9 5	13 3	5 0	14 8	10 6	13 3
December	7 2	12 2	4 1	11 4	10 3	10 8

* These abnormally high numbers are due largely to 1916, when meteors came in great numbers.

HOURS	COULVIER-GRAVIER	SCHMIDT	HOFFMEISTER
5-6	7 2	4 2	
6-7	6 5	5 3	
7-8	7 0	5 7	
8-9	6 3	6 7	5 2
9-10	7 9	7 9	6 8
10-11	8 0	9 5	7 7
11-12	9 5	11 6	8 6
12-13	10 7	14 1	10 6
13-14	13 1	16 3	11 3
14-15	16 8	17 9	12 3
15-16	15 6	18 2	12 1
16-17	13 8	18 8	12 4
17-18	13 7	14 9	15 4
18-19	13 0		

An inspection of these two tables proves at once the statements formerly given that the last half of the year and the last half of the night are, respectively, more prolific in meteors than their first halfs.

As part of these data were available to Schiaparelli in 1866, he desired to bring observation and theory into accord, but for reasons that appeared sufficient to him he concluded to investigate the number of radiants visible, rather than the number of meteors. Therefore to compare his conclusions with observations it is obvious that it was considered by him that for any given zone of the sky the number of meteors were strictly proportionate to the number of radiants He based his theory upon the assumption of uniform distribution of radiants and the parabolic velocity for the meteors, and in *Sternschnuppen,* Chapter III, developed the necessary equations from which to derive the apparent distribution of radiants over the celestial sphere. His justification for assuming the parabolic velocity was given as follows. The earth's velocity $V = U(1 - \omega)$, where ω is a small quantity never $> \frac{1}{60}$. For a meteor orbit $v = U\sqrt{\frac{2}{r} - \frac{1}{a}}$ $= U\sqrt{2}\sqrt{\frac{1}{1+\omega} - \frac{1}{2a}} \therefore v = U\sqrt{2}\left(1 - \frac{\omega}{2} - \frac{1}{4a}\right)$. Now dividing v by V we get $\frac{v}{V} = \sqrt{2}\left(1 + \frac{\omega}{2} - \frac{1}{4a}\right)$. In this $\frac{1}{4a}$ is only of appreciable influence for bodies of short periods. For the Leonids it would be only equal to $\frac{1}{41}$ for instance. Hence $\frac{v}{V} = \sqrt{2}$ approximately, and for the average case he felt that the parabolic velocity could be safely assumed. (Note: The possibility of hyperbolic orbits was not considered and it is obvious that if large numbers of hyperbolic orbits can be proved to exist, the theory would not hold.) It has been shown on page 172 that the true elongation of a meteor from the apex and the apparent elongation are very different things except at 0° and 180°, when they coincide. An abridgment of a table, due to Schiaparelli, follows on next page.

If the number in the third column be multiplied by 29261 meters/sec., which is the mean orbital velocity of the earth, the linear velocity in that unit can be found. (The last column will be very slightly erroneous, due to a solar parallax of 8.95″, instead of 8.80″, being used. For comparative purposes the tabular values are quite close enough.) It is to be understood that the last column gives the velocity of the meteor when that due to the earth's attraction is added in. We thus see that a meteor from the apex actually moves 4.34 times faster than one from the anti-apex, or neglecting the earth's attraction 5 82 times faster.

APPARENT ELONGATION	TRUE ELONGATION	RELATIVE COSMIC VELOCITY	ACCELERATED VELOCITY
0°	0° 0′	2 414	2 444
30	50 42	2 189	2 222
60	97 46	1 618	1 662
90	135 0	1 000	1 070
120	157 46	0 618	0 726
150	170 42	0 457	0 595
180	180 0	0 414	0 563

The theory is developed further as follows, assuming a great number of streams and an equal distribution. If the earth were without motion the radiants would be distributed equally over the sky. This assumes the earth to be the center of a sphere, whose radius is v = $\sqrt{2}$. Then from all parts of the surface of the sphere a certain number of meteors in a unit of time would appear. Let ASS′ (fig. 13) be this sphere, center at O, radius = v. Now give the observer a motion in direction OD with velocity V, then the results will depend upon the relative motion at O of the meteors falling with velocity v, and of the motion of the observer who arrives at O with velocity V.

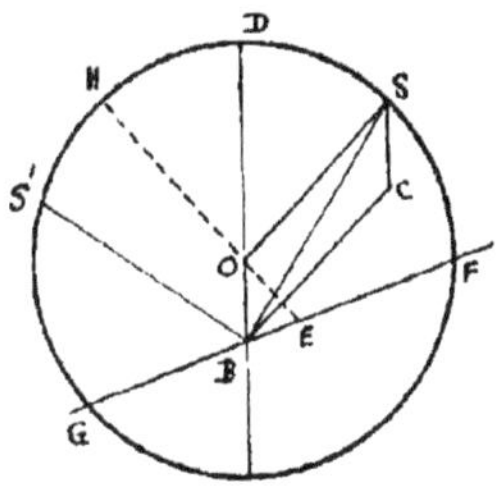

FIG 13

Then the meteor S which falls at O with velocity v = OS must have fallen along SB, resultant of OS = v and SC = V, where V is counted in the opposite direction from that in which the observer moves. As OB = SC = V, the point B always remains inside the sphere on whose surface S remains. Thus will meteor S′ fall along S′B, and so for all others. If the observer is at B, from all parts of the spherical surface and indeed in a number proportional to this part of the surface, meteors will fall not at O, but in directions SB, S′B, etc. Hence meteor showers (or radiant points) will have greatest density in direction BD, namely in that of the apex toward which the observer moves, the smallest BA in the opposite direction. Obviously the greatest and least density are to each other as BD·BA = $(v + V)^2 : (v - V)^2$. As $v = V\sqrt{2}$ very nearly, this ratio is approximately 35·1. The lines SB, S′B, then represent the relative velocities with which meteor showers coming from S, S′ etc., meet the observer. Draw the small circle upon the sphere with S for center, and draw to its circumference an infinite number of lines from O; a very narrow cone is then obtained whose apex lies at O and whose angle can be measured by means of the surface of the area mentioned. If from B similar lines be drawn they will include the same area as that seen from O, so a second cone is obtained with apex at B, whose angle will bear to that of the first cone the ratio $\frac{\cos OSB}{BS^2} : \frac{1}{OS^2}$ or as $OS^2 \cos OSB : BS^2$. As the same number of radiants are seen from O as from B, the apparent density of the radiants as seen from O and B will vary inversely as the angles under which one sees the small area. As that seen from O is the true density, so the proportion to the apparent density is $BS^2 : OS^2 \cos OSB$. We therefore conclude that the same relation holds for points distant from the apex. The following (abridged) table clearly shows the great diminution as we leave the vicinity of the apex.

APPARENT ELONGATION OF APEX DBS	ANGLE OSB	APPARENT DENSITY ÷ TRUE DENSITY = 1
0°	0° 1′	2 987·1
30	20 42	2 561
60	37 46	1 656
90	45 0	0 707
120	37 46	0 241
150	20 42	0 111
180	0 0	0 086

If the apex is taken as pole and a small circle be drawn 54° 44′ from it, one-half of the radiants would be included in this area, only 0.2113 of the whole sphere. If the sphere is equally divided as to area that half containing the apex as pole will contain 5.82 times as many radiants as the other half.

An important result is that the number of radiants visible to an observer will depend upon the height of the apex above the horizon. To find the expression let GF (fig. 13) represent the observer's horizon which obviously bounds his field of observation and let ∠DBF =

H′ be the apparent altitude of the apex. Then only those meteors will be visible which come from the part GHF of the sphere; therefore the number of radiants visible will bear to the total number of radiants the same ratio as the area of the segment GHF to the whole spherical surface. But the area of the zone GHF $= 2\pi v \cdot HE = 2\pi v(OH + OE) = 2\pi v(v + V \sin H')$. If N is the total number of radiants, and n the number visible above the horizon then

$$N = K(4v\pi), n = K(2\pi v)(v + V\sin H')$$

where K is a proportionality factor. Hence $\frac{n}{N} = \frac{1}{2}\left(1 + \frac{V}{v}\sin H'\right)$; therefore $n = \frac{N}{2}\left(1 + \frac{V}{v}\sin H'\right) = \frac{1}{2}N\left(1 + \frac{V}{v}\cos z\right)$ or

$$F = 1 + \frac{V}{v}\sin H' \qquad (1)$$

where $F = \frac{2n}{N}$. If $V = o$ then $n = \frac{1}{2}N$, or one-half of the radiants would always be visible.

The following table was calculated for parabolic velocity, i.e., $v = V\sqrt{2}$

H′ = 90° − z	F	H′ = 90° − z	F
0°	1 000	−0°	1 000
10	1 123	−10	0 877
20	1 242	−20	0 758
30	1 354	−30	0 646
40	1 455	−40	0 545
50	1 542	−50	0 458
60	1 613	−60	0 387
70	1 665	−70	0 335
80	1 697	−80	0 303
90	1 707	−90	0 293

As the apex is always in the ecliptic nowhere outside of the tropics can it ever be in the zenith. For instance in latitude $\phi = +38°$ the maximum value of H is 75½°. It is further evident that as L is always about 90° ± 1° less than ☉, on an average the meteoric apex will be on the meridian at upper culmination about 6:00 a.m. or 18 hours, and below the horizon at lower clumination about 6:00 p.m. As in

THE GLOBULAR STAR CLUSTER N.G.C. 6656 = M 22
BY J. C DUNCAN AT MT. WILSON OBSERVATORY

*

temperate latitudes twilight is strong or it is full daylight at both of these hours for all except the months near the winter solstice, the extreme maximum or minimum ratios would not be actually observed. As the equinoxes may be considered average days at these we would find the ratio about 3:1, or three times as many at dawn as at dark.

The following equation suffices to calculate H′ at any moment for a place with latitude = ϕ

$$\sin H' = \sin \phi \sin \delta + \cos \phi \cos \delta \cos (\theta - \alpha)$$

where (α, δ) are the coordinates of the apex and θ the siderial time.

This theory of daily variation is developed approximately as by Schiaparelli and rests, it must be remembered, upon two assumptions: uniform distribution of radiants and parabolic velocity for the meteors. When we attempt to apply such theory in practice we meet almost insuperable difficulties. Among them is that taking the whole year, meteors and presumably radiants are about twice as numerous in its second half as in its first. (This refers to our northern hemisphere. Too little has been found out about the distribution of meteors seen in our southern hemisphere for there to be any certainty concerning it.) Also the varying hours of sunrise and sunset and the consequent variations in the length of twilight for all places except near the equator make it impossible to test the law fully. Moonlight complicates half the hours of darkness, and meteorological conditions change greatly in the course of the work on many nights making the numbers per hour not comparable, etc The best that we could possibly expect therefore is a more approximate fulfilling of such laws even if great amounts of data are used in testing them.

The theory just given at some length is not complete, as Schiaparelli himself points out, and proves in the Sixth Note, pages 256–261, of his book. It can only be considered a first but most useful approximation because he says he neglected some of the important conditions which influence the factor a in $F = 1 + a \cos z$, the form so far given. It led him however to discover the connection between comets and meteors, even in its imperfect form. Von Niessl stated that the reason why the simple expression $1 + a \cos z$ does not satisfy the conditions is that if it could be assumed that the number of meteors for any moment were proportional to the number of radiants then this expression would be correct. But both from theory and observation this is known not to be the case, and the number of meteors from a given radiant is, other things being equal, dependent upon cosec e′ and the influence of the zenith distance. Hence the expression is more complicated.

The corrected expression is derived by Schiaparelli as follows:

The data as to the number of meteors in a given hour, found in § 70, are the means for all days of the year during which these meteors were observed. The same hour of the day corresponds to different altitudes of the apex according to the date To introduce this circumstance into the calculations, the apex may be considered to move in the ecliptic with uniform motion, at a distance behind the sun which deviates from 90° by ± 1°. If we therefore choose the zero epoch at the summer solstice, the longitude L of the apex will be almost proportional to the time and the increment of the time can be expressed by the increment dL.

In figure 14 let PZM be the meridian of the observer, S the sun, P the pole, Z the zenith, ASM the ecliptic, and A the apex. The hour-angle of the sun or the true time will be $\angle$ MPS = θ On grounds stated let SA = 90° The $\angle$ APS at the pole of the equator will in general not be 90° . by calculation its value varies from 94° 56′ to 85° 04′. It can be expressed to within 2′ by the formula $\angle$ ASP = 90° + (4° 56′) sin 2L. We denote the constant by a, then $\angle$ APZ = 90° + θ + a sin 2L Of the two sides of the Δ ZPA, PZ and PA, the first is the complement of the latitude of the observer, which we will call 90 − ϕ, the other the polar distance of the apex which we know from cos PA = sin L sin ϵ, where ϵ is the obliquity of the ecliptic In Δ ZPA the side ZA = the zenith distance of the apex, which we will call z. Then we have

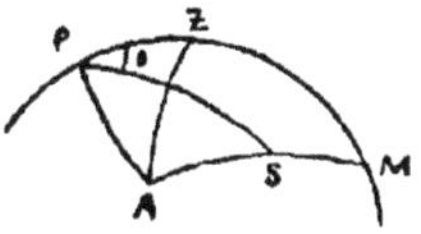

FIG. 14

$$\cos z = \sin \phi \cos PA + \cos \varphi \sin PA \cos (90° + \theta + a \sin 2L).$$

In this we substitute for cos PA its given value, and $\sin PA = \sqrt{1 - \cos^2 PA} = \sqrt{1 - \sin^2 L \sin^2 \epsilon} = 1 - \frac{1}{2} \sin^2 L \sin^2 \epsilon +$ which is exact enough, as sin ϵ is a sufficiently small fraction. Also substitute for

$$\cos (90° + \theta + a \sin 2L) \text{ its equivalent } -\sin (\theta + a \sin 2L).$$

This can be developed, considering a sin 2L as a differential,

$$\cos (90° + \theta + a \sin 2L) = -\sin \theta - a \sin 2L \cos \theta.$$

Then substituting we get:

$$\cos z = \sin \varphi \sin L \sin \epsilon - \cos \phi (1 - \tfrac{1}{2} \sin^2 L \sin^2 \epsilon) (\sin \theta + \sin^2 L \cos \theta).$$

If this value is substituted in formula (1) and multiplied by dL and integrated with respect to L between limits 0 and 2π, by which the integral serves for the entire year, and then divide the result by 2π, a number is obtained which expresses the mean frequency corresponding to θ in true time at latitude ϕ, through the formula.

$$(2) \qquad \frac{1}{2\pi}\int_0^{2\pi} dL\ (1 + A \cos z) = 1 - A \cos \phi \sin \theta\ (1 - \tfrac{1}{2} \sin^2\epsilon)$$

It must be understood that here it is no longer assumed that $A = \frac{v}{V}$, as was done by Newton, but $A = \frac{v}{V}(1 + \delta)$, where δ is an unknown function depending upon different circumstances influencing the series of observations to be used. Equation (2) expresses the mean frequency, i.e., that which is given for all epochs and all hours of the day, on an average. If K denotes the mean absolute hourly number, also the number of meteors which are observed in the mean during any given hour of the year, then the average number of those meteors, observed in an hour beginning at $\theta - 30^m$ ending at $\theta + 30^m$ (within a vanishingly small error) can be expressed by

$$K\,[1 - A \cos \varphi \sin \theta\ (1 - \tfrac{1}{2} \sin^2 \epsilon)]$$

From this expression it is seen that at the earth's poles the number is a constant but varies most at the equator . . . .

Using this formula for the series of Coulvier-Gravier and Schmidt he obtained a fairly good, but by no means excellent, agreement with each. The deviations were principally toward dawn when the meteors did not increase in numbers as fast as theory demanded. This was the final form derived by Schiaparelli, and on the whole was tolerably satisfactory for the observations extant at that date.

In 1878 von Niessl,[1] using among others the catalogue of Greg elsewhere mentioned, followed up this question. His principal reason was that observations never pointed out the maximum at 18^h expected according to the above theory. The argument is that if meteor frequencies depend upon zenith distance only then the maximum must certainly fall at 12^h, but if upon distance from the apex only then at 18^h. As neither of these is borne out by observations it is not unreasonable to try to combine the effects, since the maximum actually, according to observations, falls between the two hours mentioned. He states also that at first he thought that the very easy explanation of the effect of twilight and dawn must largely influence the fact that the maximum is not found to be at 18^h, but further research proved this not to be the case. He then set up expressions in

[1] *Astr. Nach*, **93**, 209, 1878.

which the apparent density of meteors at one and the same distance e from the apex is constant and represented by f (e), so that the density zones concur with the elongation zones. And further he assumed that each element of area furnished meteors in proportion to the cosine of its zenith distance. (In his development the influence of zenith-attraction is omitted) His final expression is, assuming the parabolic velocity and with substitution of constants in f (e), an empirical formula, where the coefficients are relative numbers:

$$M = 4.2 + 6.9 \cos z + 5.8 \cos^2 z$$

For $\phi = 45°$ he calculates M for each hour of the night, averaging the two solstices and the two equinoxes which he considers a fair mean value for the whole year. The values for 6^h, 12^h, and 18^h are respectively 2.4, 4.5, and 11.8. If one compares the values at 12^h and 18^h, as given by this computation, with the series of Coulvier-Gravier and Schmidt, the agreement is very poor. Von Niessl says this can be bettered by assuming a velocity greater than $\sqrt{2}$ for the meteors, i.e., hyperbolic velocity. Even then the agreement is not good, and he concludes that the most favorable position for a radiant, so that its meteors can be visible, is not the zenith but a zone at a certain distance therefrom. He concludes, among other things, that the theory of strong concentration of radiants near the apex does not correspond with the observed facts of daily variation. This latter would be better explained by an assumption of smaller density differences and greater velocity. Therefore the parabolic velocity does not seem to be borne out by observation, this being also shown by the inequality of the distribution of radiants which is smaller than that demanded by theory. But if a concentration is to be assumed the proper place lies between the apex and the antihelion Observational data bore out this last A natural explanation of this lies in the otherwise grounded assumption that the density of the perihelia of meteor orbits does not diminish with growing perihelion distance. And lastly that an assumption of strong hyperbolic velocity fits all theories as well as the parabolic assumption. As to this research it seems necessary to say that some of the conclusions were partly based upon Greg's Catalogue, referred to on page 85, and the criticisms of its material necessarily apply to any results drawn therefrom. Data of vastly higher quality such as Tupman's and Zezioli's were, however, also employed by von Niessl.

The next serious study which attempted to compare observations with theory was made in 1885 by General Alexis de Tillo.[2] He used the Russian Catalogue of Kleiber in which the latter calculated 1490 parabolic orbits for as many radiants. These radiants were a conglomeration of everything, good and bad, derived by nearly all previous observers of prominence, and hence had the misfortune of being far from homogeneous Yet uncertain as many of the data were, an assumption that the errors of judgment and observation of the various observers balance one another may provisionally be made. At any rate a study of these data would indicate the order of the numbers to be expected, even if unable to give the numbers themselves with any approach to accuracy. De Tillo handled the data skillfully and his results are very important and interesting. We copy one of his tables of relative numbers of radiants counting from the sun, and referring to the northern hemisphere of the sky only

0°	to	30°	0 8%	180°	to	210°	18 5%
30		60	1 4	210		240	18 0
60		90	2 3	240		270	13 9
90		120	5 2	270		300	9 8
120		150	8 4	300		330	4 5
150		180	15 2	330		360	1 5
Total.			. . . . .				100 0%

We note at first with surprise that it is the region around the antihelion rather than around the apex which gives the maximum. This is readily explainable, however, in the writer's opinion by the fact that the radiants near the antihelion may be observed equally well at all times of the year since it passes the meridian at midnight, while for many months of the year the apex crosses the meridian in daylight, and for others during twilight. De Tillo himself states that the explanation of the second half of the year giving meteors in the ratio of 100:37 compared with the first half, as found by Denning (and others), is that the apex is south of the celestial equator during the period from the winter solstice to the summer solstice, and north of it during the second half of the year. It is true that if he is correct in this, which certainly seems a partially correct explanation, we would

[2] *Bul. Astr.*, **5**, 237 and 283, 1885.

find the figures reversed in the southern hemisphere of our earth. It is certain that to date not enough observations have been made either to prove or disprove this point by persons south of the equator, and therefore for the present this explanation may partially stand. Hoffmeister's conclusions on the subject, which are the latest published, will be found on page 197.

Besides what has been said de Tillo concludes among other things that no concentration of radiants is found around ρ Herculis, i.e., the apex of the motion through space of the solar system; but that regions in the Milky Way seem much richer in radiants than others. This may be due, he says, to the position of the meteoric apex during the second half of the year north of the equator. In such a position it passes through the galactic regions at its greatest altitudes. He found the ratio in the northern hemisphere of the sky between galactic and non-galactic regions to be 1.4 to 1. A. Svedstrup[3] had found that most comets have their perihelia in the Milky Way. De Tillo found the radiant density a function of the declination, growing smaller toward the equator. He correctly explained this by the results of the presence of both the sun and the moon which on the whole light the equatorial regions of the sky more than the polar. He found a "mean functioning duration" for the radiants of 17.5 days, but states that this probably only proves that most of the 1490 radiants did not belong to different meteor streams.

De Tillo's result that no concentration of radiants was found near ρ Herculis led von Niessl in 1895 to undertake a comprehensive mathematical research into the influence of the space-motion of the solar system upon the distribution of demonstrable meteor orbits.[4] As this memoir is both very long and mostly mathematical space permits little but the conclusions to be given here. His argument is that if we attempt to consider the results of the motion of the solar system through space upon the distribution of meteor radiants in the same manner as we did that of the earth's motion around the sun, we are led into error. In other words the resultant cannot be determined by merely summing the two vectors, one the velocity of the earth in its orbit, the other the motion of the solar system in space. He reasons that the apparent radiant point of even a parabolic meteor is not the same as the true radiant. And also that this latter can be a whole

[3] *Astr. Nach.*, **107**, 113, 1883.
[4] *Denks. d Kai Akad. d. Wissen , Wien*, **62**, 449, 1895.

Coopertown, Robinson County, Tennessee, Meteorite

An etched section Photograph loaned by Geo P. Merrill Original weight 47 kilos. Composition 89 59 per cent iron, 9.12 per cent nickel

quadrant from the direction of aphelion which fixes the direction of the major axis of the orbit in space. In other words the direction of motion of the meteor on striking the earth is not the direction of its starting point in space. He also mentions (but with much less force in view of modern data) that the space-motion is less well known than the orbital motion of the earth. Therefore if all meteors came from space directly toward the sun, and all cosmical starting points were united about the solar apex, then no radiants whatever could be found in that part of the heavens. Under all reasonable assumptions of velocity, these would further be found in positions dependent upon the position of the earth in its orbit etc. This single example of his reasoning must suffice for our purposes as it proves his first statement.

He then concludes, for the case of probability equal in both directions, that the original, absolute direction of motion of cosmical meteors can have their density maximum upon a small circle, with solar apex as pole, not more than 90° distant from it. However if more move in the same direction as the sun than in the opposite, then the circle of maximum density can lie anywhere from 0° to 180° from the solar apex, and two maxima are possible. For example the maximum would then fall at the apex if the velocities were essentially greater than that of the solar system and no direction was favored. But the maximum would be inconsiderable and not provable through observations. He finally says:

> The most favorable assumptions concerning the original distributions and velocities lead in general to the result that the average density in the distribution of starting points in the hemisphere containing the solar apex will be greater than in the opposite. At the outset it remains however unproved whether this difference is large enough to be shown by existing data from observation, or indeed if it could be proved by such data. In the question under discussion only a positive, by no means a negative answer can be given on the basis of empirical researches.

These results of von Niessl made further research along these lines seem rather hopeless. Nevertheless N. Herz in 1898 undertook the computation[5] of the resultant of the two vectors, the earth's motion and the motion of the solar system, despite von Niessl's conclusion that this could not be the proper procedure. The first velocity was

[5] *Handwörtherbuch der Astronomie*, **2**, 169 et seq.

taken as 29.5 km./sec., the second as 24 km./sec., and the solar apex was assumed to be at $\alpha = 260°$, $\delta = +32°$. (The average of the more recent determinations of Boss and Campbell gives $\alpha = 271.2°$, $\delta = +30.9°$, $v = 17.77$ km /sec.) Therefore his tabular results are somewhat wrong in view of modern data, but the method needs no alteration. The reader is referred to the article if he desires to follow out the subject in detail. Herz applied his results to two cases, one assuming the solar motion to be equal to 0.8 of the 29.5 km /sec., in round numbers, the other considering it to be zero. For latitudes $\varphi = +40°$ and $\varphi = +50°$ he calculated the relative hourly numbers for each month, on both assumptions. While the agreement with Schmidt's observations, which he used for comparison, was poor in both cases, it was very much better for the solar motion equal to zero than for it equal to 0 8. On the latter assumption a strong maximum was also shown in March, contrary to observation. He therefore concluded that his research made the cosmical origin of meteors uncertain but strongly upheld their origin within the solar system.

While the method is interesting, in view of von Niessl's memoir it is doubtful if we can place confidence in the result. In comparison with the results of Herz we may compare those of Newton on page 53 and of Schiaparelli on page 57, both of which tend to the parabolic velocity for meteors, as the least possible, if not to a greater velocity.

The last important research on this general question appeared in 1922. Its author is C. Hoffmeister[6] whose work has already been so favorably mentioned. The title is *Untersuchungen zur astronomischen Theorie der Sternschnuppen* and, as the writer states, is intended to be a continuation of the work contained in Schiaparelli's book. As this memoir is an admirable work it will receive full mention.

His data consisted of 5573 meteors observed in 848 hours of work during the years 1909 to 1921 inclusive. Having been made personally, with a special object in view, these results should be accorded a far greater weight than similar ones, based upon more meteors, but less carefully made. Let us repeat that Hoffmeister was desirous of finding an accord between theory and observation for the daily and yearly variation in the numbers of meteors. We have seen that von Niessl obtained an empirical expression for the daily variation, already given on page 190, in which the coefficients are relative num-

[6] *Astron. Abhandlungen, Band* 4, Nr 5.

bers. It is obvious that such a formula can scarcely have a physical basis. Hoffmeister by a geometrical method finds that on the assumption of parabolic velocity and equal distribution the hemisphere containing the apex as pole must have seven-eighths of the whole number of radiants, that with the anti-apex only one eighth. (Compare with result on page 185.)

He then derives an expression for the relative number N in the form of a complicated integral. Certain terms must still be included to take care of the increased velocity of the meteors due to the earth's acceleration, and also the effects upon the radiants of the zenith attraction. He also calls attention to the fact that the dependence of the numbers of meteors upon cos z is not a true expression as it would be if it were merely a question of the number of meteors which would strike upon a horizontal plane. He takes as a lower limit of visibility of meteors the altitude 5° and assumes that the stratum in which meteors appear and disappear is from 60 to 140 km. above the surface of the earth. He concludes that most meteors will be seen when the radiant is 6.5° from the zenith and derives the corresponding expression

$$M = 1\,007 \cos (\zeta - 6^{\circ}5)$$

The equations and the method are both too complicated to be reproduced here but the general form is $N = \frac{1}{2c} \int \int F\ (A,\ z)\ dA\ dz$, in which c represents the meteor's velocity. Hence it can be seen that if N is observed and the integral calculated, c can at once be found; inversely if c is assumed N can be calculated. His final empirical equation for N, or inversely for c, is complicated, containing terms cos z, sin 2z, cos 2z, and powers of these functions. Yet the beauty of the equation is that nothing but constants and three variables N, c and z appear. In other words as z is always known for any desired moment and N can be observed, then c the velocity can at once be derived from the equation.

Hoffmeister combined his observations so that those made at the same zenith distance of the apex were averaged. He thus formed the following table which is of great interest as proving both the dependence of meteor frequency upon the zenith distance of the apex and its non conformity with the parabolic hypothesis.

Z	NUMBER OF UNITS COMBINED	HOURLY NUMBER METEORS	RELATIVE NUMBER		O − C
			Observed	Calculated	
37°	5	25 1	2 44	2 18	+0 26
49	5	17 1	1 68	1 94	−0 26
60	6	17 9	1 74	1 69	+0 05
70	5	12 8	1 24	1 46	−0 22
75	10	14 6	1 42	1 34	+0 06
80	11	11 1	1 08	1 22	−0 14
85	24	10 7	1 04	1 11	−0 07
90	14	9 8	0 95	1 00	−0 05
95	38	10 6	1 03	0 90	+0 13
100	19	8 7	0 85	0 80	+0 05
105	20	9 0	0 87	0 70	+0 17
110	14	6 2	0 60	0 62	−0 02
115	22	7 4	0 72	0 54	+0 18
120	16	7 0	0 68	0 47	+0 21
125	16	5 6	0 54	0 41	+0 13
130	13	6 4	0 62	0 35	+0 27
135	7	6 4	0 52	0 30	+0 22

He also combined the observations from the best three out of his four series into 15 groups with regard to z and compared N as calculated with N as observed. The accord is really wonderful, the greatest difference $N_c - N_o$ being about 7 per cent, the average only about half so large. In this case c was taken as computed, not assumed as equal to $\sqrt{2}$. He also found means to free his observations from the influence of z. Then his hourly numbers for 33 positions of the sun, throughout the whole year, were computed. These were:

☉	k	☉	k	☉	k	☉	k	☉	k	☉	k
6°	9 2	68°	6 7	139°	12 7	196°	9 4	268°	9 6	338°	9 3
16	8 0	74	6 5	149	7 3	205	11 9	282	8 1	345	8 7
25	8 8	86	6 5	155	7 6	215	11 9	297	8 7	356	9 9
38	8 3	98	6 9	165	9 2	224	12 1	306	7 4		
46	7 8	115	6 6	174	12 6	237	12 3	316	6 9		
57	6 7	126	11 6	184	8 8	253	11 1	324	7 7		

In these numbers the three streams, with elliptic orbits, the Lyrids ☉ = 31°, the Perseids ☉ = 139°, and the Geminids ☉ = 260°, were omitted. He explains the larger figures for ☉ = 126° and ☉ = 139° by the presence of the δ Aquarids, Cygnids, early Perseids, etc., for which it was not possible to make allowance. If a smooth curve is drawn through these 33 points the yearly distribution may be found. The smoothness of this curve argues well for the theories upon which it is founded. Actually he finds the ratio 2:3 for the two halves of the year instead of the ratio of 37:100 found by Denning and similar ones by others, which have already been mentioned.

Hoffmeister also studies the older series of Coulvier-Gravier and Schmidt, so frequently used by his predecessors. While from these he derives a confirmation of his own high velocities for meteors, he is only able to give inferential reasons why their frequencies fell off after 15 hours. Yet those given, which include influences of place of observation, city lights, morning fogs, dawn, etc., and again the strong influence of the much-observed Perseids whose maximum must come at approximately 15^h or earlier for dawn begins shortly thereafter, also that faint meteors were more frequently missed than by modern observers—the combination of these causes seem quite sufficient to explain the apparent anomaly. He himself does not find such a falling off in his observation, which may be found in the table on page 182 but admits a less sharp rise after 15 5^h than anticipated. He, however, notes that the number of available observations after this hour are relatively few.

It would not be out of place to call attention to the growing fatigue of an observer after a long night's work, hence more meteors would be missed in the very last hours of a long watch than during the first That this must be a very real cause, particularly in the missing of very faint objects, can be denied by no one. It seems that this is a most important consideration in this connection.

In closing it should be said that Hoffmeister's method and results deserve the closest study and ought to be tested by other series of observations. The writer eventually hopes to do this with that part of the 35,000 or more observations, made by the American Meteor Society, which on study will be found useful for this particular problem.

Very recently an important extension of Hoffmeister's work[7] has been made by B. Fessenkoff and B. Sĕigolev of Moscou. Using his table copied on page 196, they calculated the relative numbers of

[7] As this goes to press Hoffmeister publishes an important addition to his earlier work in *Astr. Nach* Nr 5302-3, 1924. He therein presents further proof of the cosmical origin of most meteors.

meteors, for $\delta = +23.5^\circ$ and $\delta = -23.5^\circ$, supposing them to move with elliptical velocity. Their results show at once a far poorer agreement with observation than is given by the assumption of either the parabolic or hyperbolic velocity. Hence they conclude that no great per cent of our meteors are derived from disintegrated comets of our system, and that their work leads them to a similar conclusion with Hoffmeister, i.e., that most meteors move with hyperbolic velocity and must therefore have originated outside the solar system.

CHAPTER XVII

The Formation of Meteor Streams

So far meteor streams have been considered simply as existing and little has been said about how they came into being. It is true that in specific cases it has been proved that a few of the larger streams follow approximately the same orbits as certain comets, and the connection is too close to be fortuitous. The contrary fact has not yet been stated, however, that some comets whose orbits come quite near the earth's seem to furnish us with no attendant meteor streams. Nevertheless this is the case. Finally it later will be shown that there is strong reason for believing many meteors come from space, while comets certainly do not but seem to be original members of the solar system. All these facts complicate the question more than seemed to be probable when Schiaparelli first discussed it.

Another point of view must be stated which may be summed up in the question which is the original, comet or meteor stream? Or better stated and with more probability, is it possible for a meteor stream to be condensed by the perturbations of planets until, in extreme cases, the densest part appears to be the head of a comet? Perhaps we have been too prone to accept the comet as being the original body, because dispersive actions are more easily calculated than formative actions, and because the former would be more easily observable from the earth. But it is very necessary in a full discussion of the phenomena never to lose sight of the possible agency of formative forces, as well as their opposites.

The final decision of the origins of the material rests upon the broader question, how was the solar system formed? Fortunately both the Nebular Hypothesis and the Planetesimal Hypothesis offer obvious chances for large amounts of material to be left over after the formation of the planets and satellites. But it does seem difficult, using the former, to explain how small solid bodies of the size of meteors could ever have come into being, unless we postulate the formation and later disruption of very many larger bodies of the order of asteroids in size. Even if that is granted, how can we explain that meteors cut our orbit at every possible angle? It is true that for this latter fact even the Planetesimal Hypothesis might be found not wholly adequate, but it has very many obvious advantages by giving us all the small, solid bodies we could need in our theories, already extant in the solar system. However, if we take the view recently expressed that the solar system, if thus formed, is rather unique and there can be few other similar ones in the whole galactic system, then we are face to face with the original difficulty, how are those meteors and meteorites formed that are proved to come to us from outer space?

While we are here able to offer no solution for this difficulty, it seems wise to state it concisely and to say that any theory of evolution for stellar systems can not be considered complete unless it adequately explains the formation of meteors, meteorites, and comets. And so far most theorists have been far too busy trying to explain the planetary part of the system to pay much attention to the smaller, but very important, meteoric and cometary members, which are present in such vast numbers.

Schiaparelli studied this question at length in the eighth chapter of his book. He first computes the stability of a spherical body, radius r, formed of small, homogeneous units separated from one another by the distances 2d, the mass of each m. Now if the sphere is at a distance R from the sun, whose mass is M, then the attraction of the sun on its center will be $\frac{M}{R^2}$. He calculates the attractive central force in the body to be $\frac{mr}{d^3}$ for a unit on its surface, and the repulsive disturbing force of the sun to be $\frac{2Mr}{R^3}$. Then if $\frac{2M}{R^3} > \frac{m}{d^3}$ the dispersion of the spherical system of small units must take place. We do not give his proof as later Charlier[1] and Picart, both, discussing the same problem analytically, found the expression should be $\frac{3M}{R^3} > \frac{m}{d^3}$. Schiaparelli, of course using his own formula, deduced that at distance unity in such a meteoric system if each unit weighed 1 gram and was distant 1.86 meters from the next it would not be dispersed. (The sun's density was taken as 1.5 in the computation.) For a similar system made up of continuous matter he finds $R^3\delta < \frac{1}{3,310,000}$, δ

[1] *Bul. Acad St Petersburg*, **32**, No 3.

being the density of water and R the astronomical unit. This gives for 10 cu. meters, 3 grams of matter, corresponding to atmospheric density at 0° C and p = 0.177 mm. He notes that the density thus obtained is far greater than that assigned to the coma of comets. When such a spherical body approaches the sun nearer than one astronomical unit at once the dispersive action will begin. A table was then calculated to find at what distance the planets can exert forces equal to that of the sun at unit distance. The table gives the reciprocals, i.e., to find the distance the values in the last column must be divided into 1.

PLANET	$\frac{1}{M}$	$\sqrt[3]{\frac{1}{M}}$
Mercury	4,316,550	162 8
Venus	412,150	74 4
Earth	354,020	70 7
Mars	2,994,800	144 2
Jupiter	1,048	10 2
Saturn	3,502	15 2
Uranus	20,900	27 6
Neptune	20,000	27 1

The values in the table are all slightly erroneous due to the value of the solar parallax employed, but for illustrative purposes are excellent. We see that even the earth could disrupt such a body at a distance of 1 ÷ 70.7 or about $1\frac{1}{4}$ million miles. This fact must be held in mind for such cases as that of Biela's Comet.

Comets, he concludes, can then be destroyed or changed into meteor streams (1) by the sun, (2) by a planet, (3) by sun and planet. In the first case if the body is denser near its center than in the outer layers, these latter will first be dispersed. Even if the latter is dense enough to escape the dispersion at the limit given by the equation, as it nears the sun the heat of the latter will act upon it and perhaps hasten its destruction. Again successive returns may complete the dispersal already begun during earlier perihelion passages. Some interesting examples for the Leonids, perihelion distance = 0.98, are then worked out to see how close they must have come to Jupiter, Saturn or Uranus for a path originally parabolic to be changed into the present 33-year ellipse. For Jupiter it is found to be 0.0132 astronomical units, for Saturn 0.00425 and for Uranus 0.00054.

Under these assumptions the dispersive effects to those of the sun will have the ratios 390 times, 3500 times, and 300,000 times respectively.

Case (3) mentioned above is that of a stream whose perihelion distance might be so changed as to then come within the critical distance of the sun. If an immense aggregation of cometary or nebulous matter, distant from the sun, came under the influence of its attraction, then those parts nearest would break off first and begin to fall toward it in orbits sensibly parabolic and having the same elements. But the further parts too would begin to move along the same approximate path, so that we would have a succession of cometary bodies, moving in almost the same orbits, reaching the sun at irregular intervals, over a period of very many years

An investigation of the aphelia of comets' orbits was undertaken by Hoek of Utrecht[2] in 1865 He found seven groups of comets which had aphelia in the same general area of the heavens. The agreement in certain cases was too close to be due to chance. C A. Young[3] mentions the case of the comets of 1668, 1843, 1880, and 1882, all of which almost graze the sun's surface at perihelion and which apparently move in ellipses of a minimum period of several hundred years. These cases bear out Schiaparelli's investigations, so far as they go.

Next will be given a partial list of comets, followed by some remarks, which broke up while under observation, at least in part, forming thus excellent illustrations of the beginning of the dispersive process. The comet of 1618 (second comet), 1652, Biela's Comet, the great comet of 1882, 1889 V and 1892 V Among comets with double or secondary nuclei are reported. Donati's Comet, 1853 I, 1868 II, 1862 III, and Halley's Comet in 1910.

According to Young the nucleus of the great comet of 1882 broke up into 6 or 8 parts, which stretched finally 100,000 miles, and remained visible until the comet disappeared. A number of faint, cometary objects (at least 3 or 4) accompanied the head, but wholly detached from it. Biela's Comet has already been discussed. Barnard[4] saw not less than 5 components of the 1889 comet with the great Lick refractor. For Halley's Comet some observers reported a secondary nucleus on certain dates while a curved shape for the main

[2] *Monthly Not*, *R.A.S*, **25**, 243 and **26**, 1, 1865
[3] *General Astronomy*, 459
[4] *Astr. Nach.*, **122**, 267, 1889.

nucleus was most clearly shown on many photographs, as well as seen visually. The apparent size of the nucleus varied very greatly on different dates, its shape and the sharpness of its outlines changed.

As examples of comets which were periodic, yet after a single appearance could be found no more, may be quoted Comet 1743 I, 1766 II, 1783 and 1819 IV all with periods less than seven years, and Comet 1846 VI with a period of 13 years. It needs little imagination to see how easily such disappearances could be fitted into the theory under discussion. It is not necessary to pause to quote examples of the expulsion of matter from comets, which travels out along the tail, for this has no direct bearing upon the formation of meteor streams. It is mentioned only to show in how many ways a comet is subject to dissolution. Nor has it seemed necessary to quote at length the numerous accounts of the breaking up of the nucleus in the cases mentioned, because the fact itself is all we here desire to stress.

Returning to the development of the theories, Faye advanced another in 1867.[5] He felt that the cosmic currents postulated by Schiaparelli had never been actually observed when approaching, which he believed should be the case if they were numerous, and also felt that the inclusion of a comet, which seemed to have a different physical constitution from the rest of the stream could not well be explained. The difference between a comet's constitution and that of the rest of the stream seemed too great. His own idea was that meteors, the zodiacal light, the corona, and the resisting medium were all due to nuclear emissions from comets, especially the periodic. These must take place mostly near their perihelia; thence such material would be denser near the sun. This material would be found largely in the zone near the ecliptic, as most short period comets have small inclinations.

Schiaparelli, accepting Bessel's proof that a repulsive force is the cause of comets' tails and other emissions, nevertheless refused to believe that material driven back along the radius vector from the sun can form stable meteor streams He shows that they would have no element in common with the comet, except inclination and node, and that the matter would (mostly) be driven away in hyperbolic orbits. As he had also proved for the Perseids and the Leonids that the meteors followed the same approximate orbits as the two associated comets, he felt fully justified in denying the possibility of meteor

[5] *Comptes Rendus*, **64**, 552, 1867.

streams being thus formed, even if for argument's sake the repulsive force was strong enough to drive back solid particles of the necessary size to become meteors. He was thus strongly brought back to his own original ideas that the material was scattered along the orbit, one particle following in general about the same path as the rest.

Nevertheless the original theory of Schiaparelli about cosmical clouds was at once shown untenable by Weiss[6] and was so partly admitted by its originator. In this splendid article, already quoted, Weiss shows that from the researches of A. Herschel on the mass, and those of others upon the probable distance between members in even the great Leonid stream, their distance being several hundred kilometres, only the most impossibly small density could be allowed for a cosmical current. So small in fact that the attractions of the stars would disperse it out in space and it could never even begin to move as a whole toward our system. Comets could never form parts of such a cloud but are directly the bodies whose disintegration produce meteor streams, and short-period elliptical comets will be those most exposed to the solar action and hence will most quickly disintegrate. Further he stated that any detached meteors will follow orbits like that of the comet at the moment of detachment, not that followed by the comet later on, when it had perhaps itself meantime been greatly perturbed.

Schiaparelli in his later opinions on this subject, developed in his book § 101 et seq., accepts part of this criticism and infers that comets do not come from the regions of the stars, but while coming from distances vastly greater than planetary, yet accompany the sun on its journey through space and hence are members of our system. In 1910 the same writer again published[7] what was probably his final opinions upon cosmical currents, and which later will be mentioned more fully.

Schulhof[8] states that the theory of Schiaparelli as modified by Weiss fitted the facts fairly well but that there were the following serious objections, which are given in his own words:

1. Of some thousands of swarms which in truth are not all independent the one from the other only four are known whose connection with a comet is incontestable.

---

[6] *Sitz. d Kai. Akad. Wien.*, **57**, 2, 281, 1868.

[7] *Bul Astr*, **27**, 194, 1910

[8] *Sur les Étoiles Filantes* **47**, 1894

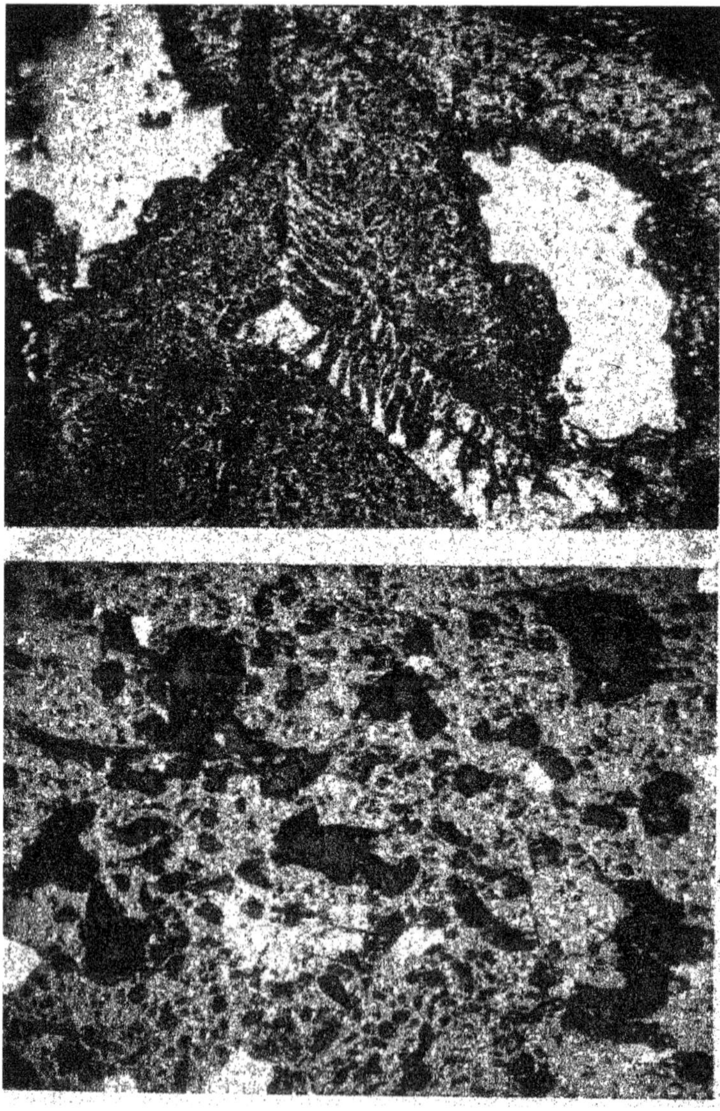

A

B

PHOTOMICROGRAPHS OF METEORITES BY L. F. JEWELL

Taken by specular reflection. From Journal of Optical Society of America.
A, Carleton, Texas, Meteorite. *B*, Four Corners, N. M., Meteorite.

2. The very number of these streams which requires the conception of an inconceivable number of distinct meteoric currents in the planetary system, admitting that one can imagine no valuable reason that they should be particularly condensed about the terrestrial orbit.

3. The long duration of certain radiants (according to Greg one must suppose these on an average as three weeks); besides radiants active only one or two days there are others active six weeks.

4. The successive displacement of radiants in the swarms of long duration.

5. The existence of radiants fixed during several weeks, even months.

The writer desires to attempt to partly answer these, in view of modern data.

1. Six of the larger streams are now known to belong to comets; a fairly satisfactory accord for several other smaller streams has been announced at intervals.

2. If we assume that three new comets (not an unreasonable number), discovered and undiscovered, come to perihelion yearly at a distance not greater than one astronomical unit, and suffer, as they must, partial dissipation or dispersion, in the millions (?) of years during which this has happened there surely would be enough material left behind to furnish our present meteoric phenomena. Nor is the writer willing to admit that there is not sound reason for there being more meteors in the zone near the ecliptic than in other zones, because the earth's orbit lies nearly in the invariable plane of the solar system and hence will profit by the perturbations of the outer planets, which have tended to draw meteors into the planes of their own orbits. The explanation given in Moulton's *Celestial Mechanics*, §163, might also give an added reason.

3. It is perfectly true that some explanation must be found to account for the widening out of a stream like the Perseids, Geminids, or Orionids to such an extent that the earth needs several weeks to pass through them. (The last two cases mentioned require about ten and twenty days respectively.)

4. It has been proved by the work of Kleiber, von Niessl and others that in such a case the radiant would indeed shift its place from day to day.

5. The question of "stationary radiants" has been too fully discussed to be taken up here. It is only desired to repeat that ready explanations, as given in the proper place, are at hand for most such cases—at least in the writer's opinion.

It therefore appears that these objections of Schulhof, except the third, are hardly valid, all the more strange as they are about the only weak point in an article otherwise one of the most valuable ever published on the subject. A graver objection, however, is given by him in a foot-note, namely that it must be demonstrated, for all those streams with retrograde motions and which therefore are far less liable to perturbations by the earth, that they approach near enough to some major planet to have undergone serious perturbations, if we are to have a great width to such a stream, following of course the general idea of Schiaparelli as to their formation.

Lists of possible accords of meteor radiants with comet radiants, which when proved lead to sound belief that the one is the product of the other, have been published by Weiss in 1868,[9] by A. Herschel in 1878,[10] and have been brought up to date by M. Davidson in 1920.[11] The first two mentioned are partly quoted and fully discussed by Herz.[12] Schulhof points out, however, with justice that mere approximate accord of a meteoric and cometary radiant certainly sometimes will be due to mere chance. This is especially true if we consider as radiants large ill-defined areas, and with centers hard accurately to determine. To evade such errors it is necessary to have some means of finding whether the comet's orbit and the orbit computed for the meteor radiant, which provisionally may be thought to be formed by débris of the former, were ever coincident. We already have discussed how these two orbits would not continue indefinitely the same, due to planetary perturbations.

Tisserand had worked out a criterion by which it generally could be proved whether two comets, seen at an interval of years, were actually the same comet, even though the two orbits differed considerably. This is:

$$\frac{1}{\alpha_1} + 2\sqrt{\alpha_1 (1 - e_1^2)} \cos i_1 = \frac{1}{\alpha_2} + 2\sqrt{\alpha_2 (1 - e_2^2)} \cos i_2)$$

Callandreau[13] using this obtained an equation of condition which must be satisfied in the case of two meteor streams, when the date of appearance, as well as the radiants are different. The combination

$$J = \frac{1}{\alpha} + \frac{2}{\alpha'}\sqrt{\frac{p}{\alpha'}} \cos i$$

[9] *Sitz. d. Kai Akad. Wien*, **57**, 2, 281, 1868

[10] *Monthly Not*, *R.A.S.*, **38**, 369, 1878.

[11] *Monthly Not*, *R A.S*, **80**, 739, 1920.

[12] *Handworterbuch der Astron*, **2**, 212 et seq

[13] *Comptes Rendus*, **112**, 1304, 1891.

has a constant value for the primitive comet and any one whatever of the smaller comets, where α' is the mean distance of the disturbing planet, and α = semi-major axis, p = semi-parameter, and i = inclination of the orbit. This when substituted in the following gives the condition.

$$0 = \left\{\left(1+\frac{2}{\alpha'^{\frac{3}{2}}}-J\right)[\sin^2\beta+\cos^2\beta\sin^2(\lambda-\odot)]+1-p\right\}^2-4\cos^2\beta\sin^2(\lambda-\odot)$$

$$\times\left\{\left(1-\frac{1}{\alpha'^{\frac{3}{2}}}\right)(1-p)+\left(1+\frac{2}{\alpha'^{\frac{3}{2}}}-J\right)\right\}\left\{1-\left(1-\frac{1}{\alpha'^{\frac{3}{2}}}\right)[\sin^2\beta+\cos^2\beta\sin^2(\lambda-\odot)]\right\}$$

where $(\lambda_1\ \beta_1)$ are the coordinates of the radiant and $\odot$ = longitude of the sun. If this equation is satisfied then the different radiants belong to the same family, the perturbing planet being at a mean distance α from the sun. While the equation certainly is a long one, still for the earth $\alpha = 1$, hence it becomes much simplified in the only practical case. As no case is recalled in which it was actually employed, except by its originator who proved that the motion of the Perseids for six weeks was well represented by the formula, it has seemed well to place it here where it again would be called to the attention of meteor observers, many of whom doubtless are unaware of its existence.

Of all astronomers Brédikhine has done most work upon the formation of meteor streams and in what ways they could be derived from comets. His numerous publications would form several volumes and a detailed review is impossible. References have already been given in the chapters on stationary radiants, about which his discussion is the most thorough ever attempted. The more important articles are written in French though appearing in Russian publications, the earlier mostly in the *Annales de l'Observatoire de Moscou*, the later mostly in the *Bulletins de l'Académie Impérial de St. Petersbourg.* His theory was given first in 1877 and was based upon the fact that comets ejected anomalous tails or jets toward the sun which however were too heavy to then be repulsed by the sun in the direction of and with the material of the normal tails. He conceived that these particles once detached from the coma would follow independent orbits, with a velocity which would be a resultant of the original motion of the comet plus the various repulsive forces in action upon each particle. Even the planes of such orbits frequently would not coincide with the plane of the parent comet. He also correctly stated that the fact that such emissions are not seen in many comets is no proof that they are not occurring, it only proves that they are not great enough to be seen at such a distance. As the emissions need not, in fact scarcely would, all occur at once, those occurring farthest from the sun would move off with one velocity, those nearer with other velocities. In this way he conceived their orbits might be, for an original parabolic comet, all sorts of ellipses and even hyperbolas.

He then finds a ready explanation for such a group as that of the comets of 1668, 1843, 1880, and 1882, they all having been emitted from a great parent comet. A periodic comet would at each return thus emit fresh material, and each part or, as we may say, little meteor stream would follow a separate orbit. These orbits together would form a bundle (faisceau), some of which could cut the earth's orbit, even if the planet could not. (For such a case see the connection between the η Aquarid meteors and Halley's Comet, taken up elsewhere at length.) Also these numerous streams would in many cases have different periods. Further the units would not move in orbits absolutely parallel, hence the phenomenon of a radiant area, rather than a radiant point, is to be expected for such cases. In course of time we would get a sort of ring or closed hollow cylinder within which (most of) the orbits would be contained. Such a ring might have considerable dimensions not only parallel but also perpendicular to the plane of the ecliptic, and on the date on which the earth pierced its densest portion, the maximum of that shower would occur. The existence of so-called secondary swarms and of radiants of long duration (i.e., the Perseids) could thus apparently be explained.

Such in brief is the original theory. But with time its originator became convinced of the important rôle played also by the sun and the planets, as already postulated by others. Therefore his later opinions are condensed as follows.[14]

> The disintegration of a comet into shooting stars (i e , meteors) can be produced: (1) By the ordinary attraction of the sun; (2) by the attraction of the major planets; and (3) by the nuclear emissions occasioned within the comet by its approach to the sun.

He adds in explanation:

> The last agent, i.e., (3), at a certain distance from the sun is more frequent than the second, more universal with comets. The second manifests itself

[14] *Bul. de l'Acad. Imp. des Sciences de St. Petersbourg,* **17**, 181, 1902.

only at short distances from the major planets, within their spheres of activity. The actions of the two agents do not exclude one another. As to the first its action is combined always with that of the second, to such a degree that it is difficult or rather impossible to evaluate quantitatively their mutual relation; it is clear, in any case, that the intensity of each of them must vary with the change of the distance between the comet and the sun. The agents (2) and (3) can, one after the other, take part in the disintegration of the same comet. This must be the case for Biela's Comet

Before going into more details it may be said that Schulhof remarks that this theory is an indispensable complement to that which attributes the origin of meteors to the disintegration of comets (as originally given by others). Heiz, however, in his very complete memoir on meteors dismisses it unfavorably in a few words. Hoffmeister in his valuable paper speaks of his work as "an essential advance," and evidently feels that it is sound. Hence there is no doubt that the work of Brédikhine was of great and permanent value, and deserves full attention.

Following[15] further into Brédikhine's theory, let us examine what will happen to the major axis of a particle, suddenly ejected from a comet toward the sun. Let j = the amount of the initial impulse, H and H′ the velocities of the comet and detached mass at the instant the impulse started, by a and a′ their semi-major axes, by J the angle made by the direction of the impulse to the radius vector r (positive behind, negative in front of r), and by p the angle which r makes with the tangent of the primitive orbit, then:

$$H^2 = \frac{2}{r} - \frac{1}{a} \qquad H'^2 = H^2 + j^2 - 2H\gamma \cos(\beta - J) = \frac{2}{r} - \frac{1}{a}$$

for the parabola $\beta = 90° - \frac{v}{2}$ (v = true anomaly)

for the ellipse $\beta = 180° - (\sigma + v)$, where $\tan\sigma = \cot E\sqrt{1 - e^2}$

If now we examine the formula for the parabola (that for the ellipse will be nearly the same except that the extreme variations in the period are in narrower limits), we find that H′ is a minimum for $\frac{v}{2} + J = 90°$, which gives the shortest period; for $\frac{v}{2} + J = 0°$ the

[15] *Sur l'Origine des Étoiles Filantes*, 1888, *from Bulletin de la Société Imp. Nat de Moscou*

orbit is hyperbolic. Supposing it varies between ± 45°, we see that before perihelion only a few of the detached masses can have elliptical orbits, after perihelion there will be no hyperbolic orbits except for negative values of J. So for the ellipse the period will be long or short in general according to whether the emission took place before or after perihelion passage.

In particular for j assumed = 0.1 and 0.2 respectively, for various values of J, r, and v, the elements a, q, and P are calculated. These two values of j are taken small to be on the safe side. It is needless here to quote numerical results but from them he concludes that for each point of the orbit of the nucleus, within indicated limits, we obtain in the plane of the orbit a whole series of elliptical orbits crossing at this point. These orbits are disposed around and are near to the original orbit and have different periods. As the eruption did not take place so that the particles became scattered only in a sector, but were driven out in shape of a cone, the particles have planes of all inclinations compatible with the chosen values of J and j. All of these planes must, however, obviously contain the point at which the emission took place. For another point of eruption upon the nucleus' orbit similar results would be obtained, and in the end the bundle or ring formed for each erupting point would together form the entire ring or meteor stream as it is known by observation. Calculations are then made which prove the thickness as well as the width of such a ring to be considerable. A great difference appears between the crossing of orbits formed after and before perihelion passage, other things remaining the same. For the first case the orbits of the faisceau diverge from a point, in the second the orbits cut the original orbit plane not in a point but in a line, which indeed gives the width of the faisceau. It is then shown how actually some radiant areas are circular, some elliptical. The above discussion presupposed a comet in a parabolic orbit; however for one in an elliptical orbit the same results would follow only in a less marked degree. Finally it is pointed out that the size of the particles, as well as j, would have a good deal to do with the orbit followed by these particles, the smaller being ejected further than the larger. The statement is also made which shows the importance of the new theory, that, were the comet completely broken up by the action of the sun alone, the meteors would indeed be spread out considerably in the plane of the original orbit, but the stream would be so thin that the earth would pass through it in a few minutes. It scarcely need be pointed out that nuclear emissions are never assumed to take place very far (2 astronomical units perhaps as a limit) from the sun. Then, for such a comet as that of 1882, we might expect the maximum number and effects, but for one with $q > 1$ we could hardly expect to see much if any evidence of nuclear ejections.

CHAPTER XVIII

THE PERTURBATIONS OF METEOR ORBITS

The elements of this most important question were so excellently set forth by Schiaparelli that no better course can be followed than to translate almost word for word certain sections of his book. The following sections refer to numbers in his *"Sternschnuppen."*

§ 79 When the earth cuts through a meteor stream it leaves in it an empty space the axis of which follows the relative direction of motion of the meteors (fig. 15). This space is however not absolutely empty, owing to the attraction of the earth It contains meteors, but in smaller numbers than the rest of the stream and its diameter is greater than that of the earth. In figure 16 an

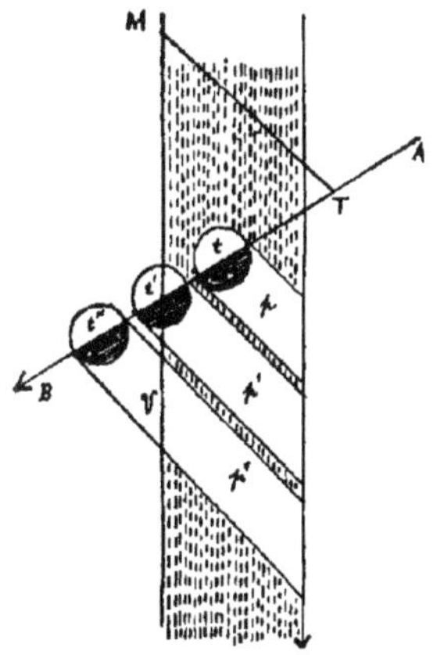

FIG. 15

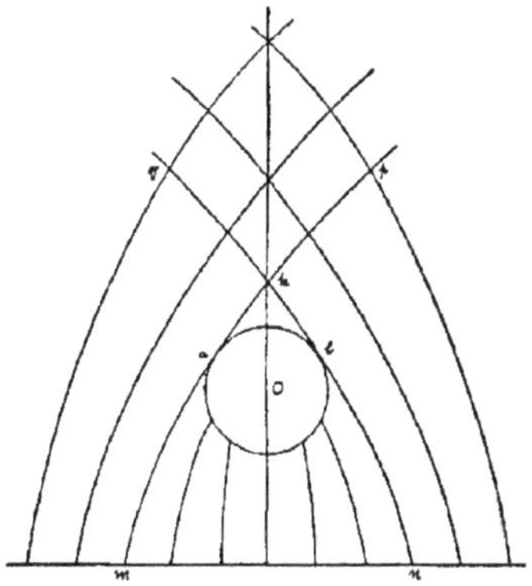

FIG. 16

exact representation of the case is given. The meteors contained in the conoid mkn will be taken up by the earth upon which they all will fall with the same velocity. those whose orbits form the surface of the conoid itself will continue their way along the after branch of the hyperbola ap, bq; finally the meteors, remaining outside the conoid, which pass by near the earth, will describe with respect to the latter just so many hyperbolas with the focus O, and will be deviated more or less from their orbits At a distance of only a few of the earth's diameters this deviation will be almost unnoticeable. The earth too, without leaving in the stream a completely empty space, distributes the meteors along a line which lies in the direction of their relative motion (or in the opposite direction of their radiant point corrected for zenith attraction and diurnal aberration).

§ 80 Any planet at O (fig 17) whose mass (in terms of the earth as unity) is denoted by μ, will, at a distance ρ, exert an attraction $= \mu g \frac{r^2}{\rho^2}$, where as usual g denotes the attractive force corresponding to a distance r from the center Now let U be the relative, undisturbed velocity of a meteor, W the velocity corresponding to the point M (for which OM $= \rho$), then we obtain for any given point M of the hyperbolic orbit the following relation, analogous to that of § 52,

$$W^2 = U^2 + 2 \mu g \frac{r^2}{\rho} \tag{1}$$

In particular let M denote the perigee of the meteor and ρ the distance of the perigee, also the least distance between the two bodies, then (1) gives a connec-

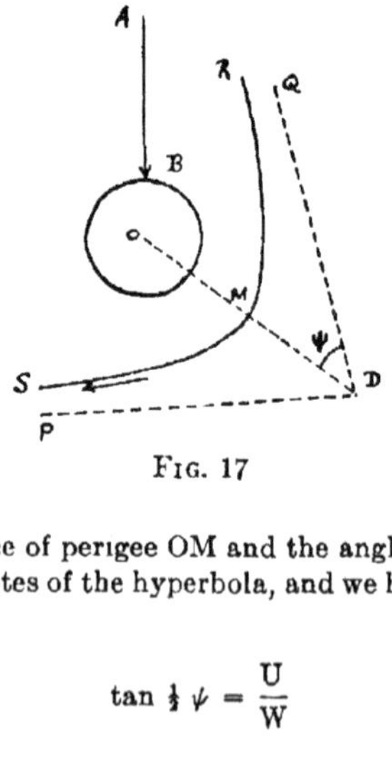

FIG. 17

tion between the distance of perigee OM and the angle 2ψ = QDP, which will be found by the asymptotes of the hyperbola, and we have the following useful relation

$$\tan \tfrac{1}{2} \psi = \frac{U}{W} \tag{2}$$

The deviation which the planet O brings about in respect to the direction of the relative motion of the meteors, obviously will be expressed through the angle PDN $= 180° - 2\psi$, formulae (1) and (2) therefore afford a very simple means of calculating the deviation of the motion when the distance of perigee is given, and this last when the deviation is given. This asserts, to be exact, that the branches of the hyperbolas are parallel to the asymptotes at those points at which the meteor enters the attraction sphere of the planet and withdraws from the same. Since the attraction sphere extends out only to a radius of about ten diameters of the planet, this hypothesis comes so near the truth that the small difference may be neglected.

§ 81. Since the hyperbolic velocity is symmetrical with respect to the perigee the relative velocity of a meteor at the point of its entrance into the attraction sphere and its departure therefrom must be the same, one further can assume that it is equal to the relative velocity U, of the planet and of the meteors, as that would be if both bodies did not attract one another Hence we also may conclude that the attraction of a planet which passes by a meteor at a short distance, brings about no change in the velocity of their relative motion, but only a change in the direction of this motion. This change, as easily seen, will be the greater, the smaller the undisturbed velocity U, and the nearer the two bodies will be to one another at the moment at which the meteor passes its perigee

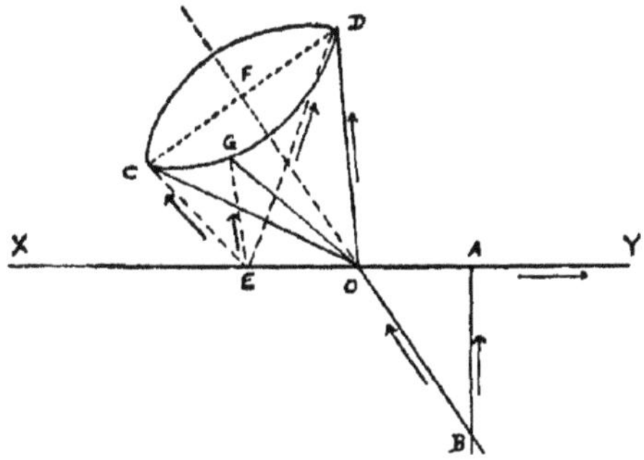

FIG 18

§ 82 Let one consider the relative motions, then the result will be different (fig 18). Let XY be the orbit described by the planet in space; we can assume for the entire time in which the meteor is found within the attraction-sphere of the planet that its motion in this path is rectilinear and uniform Further let OA be the velocity of this motion and let BA be the direction and amount of the absolute, undisturbed velocity of the meteor, the latter then will seem to fall upon the assumed planet at O in the direction and with the relative velocity BO At the passage of the meteor through perigee it will suffer a certain deviation, due to the attraction of the planet, whose amount depends upon the distance of the perigee; also instead of continuing in its path OF with respect to the planet, it will follow, without changing its relative velocity, the direction of one of the elements of the right cone DOC, whose angle at the apex FOC = FOD is equal to the above deviation Let OG be the new direction of the relative motion In order now to obtain the direction of the absolute motion, which comes after the cessation of the influence of the planet, it will suffice to add together the new relative velocity and the velocity of the planet. We also make OE = OA, OG = OB and draw EG. These lines represent the magnitude and direction of the absolute velocity of the meteor after its exit from the attraction-sphere of the disturbing planet and the direction of motion will be that from E to G.

§ 83. We further recognize that the planet and meteor are attracted by the sun. Hence BA represents the orbital velocity of the meteor in the second during which it enters into the attraction sphere, and EG the velocity in the new orbit after conclusion of the perturbation by the planet. We may assume the dimensions of the attraction-sphere as being vanishingly small in respect to the distance to the sun, also very great in respect to the distance of the meteor at perigee. In this way we are relieved of any hypothesis concerning the radius of this sphere, through which the study of all these problems would be considerably complicated, and which would hinder the development of general conclusions. For our suppositions (which in practice are assumed for each planet on account of its very small mass in comparison with the sun's mass) we may take for BA the orbital velocity, which corresponds to the origin or the point of greatest proximity of the two orbits, and regard EG as the new orbital velocity corresponding to the same point. The study of the perturbations, by which a planet can alter the shape of the orbits of meteors, when these approach it without actually striking it, will henceforth be a simple problem without the conclusions differing appreciably from the truth. Our method is essentially a simplification of that developed by La Place for analogous cases and applied by Burckhardt and Le Verrier to the case of Lexell's Comet.

§ 84. Instead of one single meteor we now will consider an entire stream, which meets the planet O with the relative velocity OB = U. Some of the elements of this stream will remain unaltered by the planet, the others will however undergo a more or less great change, the largest changes will occur for those meteors which go by nearest to the surface of the planet. Let R be the radius of the planet considered spherical, then, retaining the equation already employed, the greatest deviation D, as mentioned, can be derived from:

$$(3) \qquad \begin{cases} W^2 = U^2 + 2\,\mu\,g\,\dfrac{r^2}{R} \\ \tan \frac{1}{2}\,\psi = \dfrac{U}{W},\ D = 180^\circ - 2\,\psi \end{cases}$$

After the perturbations all meteors, which went by near the surface of the planet, will assume for their relative motions the directions of the generating lines of the cone COD, whose angle at the apex is FOD = D. The other meteors which passed at greater distances from the planet would obviously have undergone a smaller deviation, hence their relative motion after the perturbation will follow the direction of a line to be found within this cone. Hence it follows that the direction of the absolute motion after the perturbation will be contained in the slanting cone CED, which has its base in common with the other cone and for apex the point E fixed through OE = OA. The absolute velocities which before the perturbation can be assumed as equal among themselves, will then become unequal and can vary between certain fixed values.

§ 85 When we consider the case especially interesting for us, in which the earth is the disturbing planet, the angle D is the greatest deviation for a given meteor stream and is equal to twice the greatest value of the zenith-attraction which would correspond to this stream; of this one can easily satisfy himself The angle at the apex FOD will be acute and in general vary between 34° 40′ and 1° 24′ (§ 54). The greatest and least absolute velocities after the perturbation (also the greatest and least major axes of the new orbit followed by the meteors) correspond as easily is seen to the same two generating lines of the conical surface which lie in the plane of the triangle AOB, we also make OC = OD = OB then EC and ED will be these extreme values This calculation from the given OA and OB depends only upon the solution of the triangles OAB, EOC, EOD, in connection with the employment of formula (3), in which for the present case $\mu = 1$, $R = r$.

§ 86 When I worked out the earth's attraction upon the November meteors using such simple principles, I found that the angle of the greatest deviation of the relative motion for meteors grazing the earth is 1° 28′. If we indeed assume the period of these meteors as 33 25 years, then in that extreme case the perturbation can shorten the period to 28 67 years, or lengthen it to 49 92 years. The result of the passage of the meteors near the earth can also, as we see, be shown by a very considerable change in the periodic time. Although, therefore, the densest cloud of Leonids is not very long and can only meet us each time after 33¼ years, yet a small part of it, namely, the same which at one of the periods went by near the earth has the opportunity of varying its period and going into the state of a ring-like stream completely occupying the whole orbit. Hence it probably happens that, while great showers are so rare, yet almost every year traces of the radiation of the Leonids show themselves

§ 87 On the assumption that one of these meteors on its passage near the earth and upon its further progress in its relative motion has undergone the maximum deviation of 1° 28′, if this same meteor after one or two periods again falls in with the earth, as one easily sees, its radiant point will be distant 1° 28′ from the other meteors, fully understanding that its position has been corrected for zenith attraction and diurnal aberration If one further assumes that by some special chance the meteor on its second approach to the earth should graze its atmosphere and pass on, and that after one or more periods should again meet the earth, the radiant will then be observed in a third position which is distant 1° 28′ from the second and at least 2° 56′ from the radiant of the other meteors

Notwithstanding the really radical changes in the periods, the radiant of the perturbed November meteors cannot be moved a considerable distance from the main swarm Also a very great number of periods are necessary for a meteor to pass many times near the earth so that its radiant should undergo each time a considerable displacement

§ 88 This leads to the conclusion that the exactness of the phenomenon of radiation with its continuous diminution in the course of years can yet be very great, however the paths of single meteors have been widely separated from one another and at the same time have undergone considerable variations in their periods and eccentricities. A meteor stream can radiate with almost geometrical exactness from a point and yet its units describe very different orbits in space. This phenomenon must occur especially for those streams whose motion is almost exactly parabolic The attraction of the earth will then change a part of the orbits into ellipses of very long periods, some into ellipses of short periods, and others into hyperbolas. Thus the stream will be dissipated more and more, while it loses part of its members to stellar space. Those meteors however which are turned into an orbit of short and, also, more permanent period, can complete a very large number of revolutions without bing directed into a hyperbola or parabola. So for example the Leonids will always form a very stable system and none of the meteors belonging to them can be thrown by the earth into an open orbit, except after many very near passages by our planet.

§ 89. Streams coming from radiants near the anti-apex, nearly following the earth in the direction of their motion, have much less stable radiants on account of the great values which the deviation can reach—34° 40′ for those meteors that exactly follow the earth. Under these conditions, the earth can produce a maximum displacement of this amount upon the radiant Yet it must be remembered that for such a stream the number of meteors deviated from their orbits is incomparably greater than for a stream, like the Leonids, which comes almost from the direction of the apex. Therefore streams whose radiants culminate in the evening give more occasion than the others for the apparently sporadic meteors These are also the streams on which the earth produces essential changes in the major axes and the periods

§ 90. With the view of inquiring to what degree the attractive force of the earth can vary the major axes of the meteor orbits, I have worked out the circumstances under which a body endowed with parabolic velocity must move so that it might be brought, through the perturbing action of the earth, into an orbit with the least possible period On the assumption that the earth moves with its mean velocity, I find that the motion must be direct and the parabola must cut the earth's orbit, assumed circular, at an angle of 18° In this case the body describing the parabola must overtake the earth before its passage and graze the outer stratum of our atmosphere (fig 18). It will then pass out from the attraction sphere of our planet, and follow a direction inclined 25° to the earth's orbit, and describe an ellipse about the sun whose semi-major axis is 2.65 and whose period is 4 31 years. This is the smallest ellipse into which the attraction of the earth can turn a body, moving originally with parabolic velocity, which has passed by the earth only a single time. The mass of this body must be considered zero, as is self evident, with respect to that of the earth

If we place the distance of perigee, all other things remaining the same, as equal to twice the earth's radius, then the semi-major axis of the new orbit will be 5.04 and the period 11 31 years. In the following table are given the semi-major axes and periods of the new orbits, corresponding to the different distances of perigee, if the above named assumptions are always held

DISTANCE OF PERIGEE IN RADII OF EARTH	SEMI-MAJOR AXIS OF NEW ORBIT	NEW PERIOD
1	2 65	4 31
2	5 04	11 31
3	7 43	20 26
4	9 90	31 15
5	12 46	43 98
6	14 77	56 76
7	17 19	71 27
8	19 64	87 04
9	22 08	103 75
10	24 45	120 90

If we assume that the parabola is followed by a meteor stream instead of by a single body then some individual members will be deviated into orbits of short periods, as explained, others into orbits of longer period, and a considerable part (we may say half) into hyperbolic orbits We also must note how different are the results which follow the passage of the earth through the same meteor stream

§ 91 The results following the perturbations by other planets upon meteor streams passing very near them are quite similar to those produced by the earth, differing only in the degree of their intensity. The inner planets produce less effect than the earth both because of their smaller mass and because the relative velocity of the meteor when passing them is greater under equal conditions On the contrary, for the major planets, particularly Jupiter, the action of the greater mass will be reinforced by the smaller relative velocity These deviations can therefore attain values far in excess of those caused by the inferior planets. To gain a comprehensive view of the amount of deviation produced by different planets upon meteor paths, the following table, in two parts, has been computed.

In the first part a meteor stream is assumed which, following in its parabolic motion the same direction in which the planet moves, reaches that planet in such a manner that for an observer thereon it appears to give a radiant at a point opposite to the apex There is then given the relative motion (which is also for the parabolic stream the least possible), the accelerated velocity, the zenith attraction at the horizon (which is the greatest possible), and the deviation suffered by those members of the stream, which pass close by the surface (which again is the maximum deviation producible by the planet itself). In the last column the limiting distance is given (in equatorial radii of the earth) at which a meteor of the stream under consideration can pass by without its orbit being deviated more than 4°. The planets are assumed to have circular orbits and uniform motion

In the second part of the table a meteor stream is assumed whose parabolic motion is exactly opposite in direction to the planet's and which therefore produces a meteor shower from the apex Here too are listed the relative velocity (now a maximum), the accelerated velocity, the zenith attraction at

PLANET	RADIUS OF PLANET EARTH = 1	MASS OF PLANET EARTH = 1	PART 1 STREAM FROM DIRECTION OPPOSITE TO APEX					PART 2 STREAM FROM DIRECTION OF APEX				
			Relative velocity, meters per second	Velocity, meters per second	Zenith attraction at horizon	Greatest deviation	Distance corresponding to a deviation of 4° (in earth's radii)	Relative velocity, meters per second	Velocity, meters per second	Zenith attraction at horizon	Deviation for meters grazing the planet	Distance corresponding to a deviation of 4° (in earth's radii)
Mercury	0 390	0 08	19,483	20,129	1° 52′	3° 44′		113,558	113,671	0° 3′	0° 7′	
Venus	0 969	0 86	14,432	17,717	12 22	24 43	7 30	83,081	83,743	0 35	1 10	
Earth	1 000	1 00	12,120	16,482	17 20	34 40	11 74	70,642	71,520	0 42	1 24	
Mars	0 545	0 12	9,818	11,129	7 10	14 20	2 15	57,224	57,465	0 12	0 29	
Jupiter	11 640	338 00	5,314	60,419	79 56	159 51	20,651	30,971	67,686	40 48	81 37	608
Saturn	10 010	101 00	3 924	35.694	77 27	154 54	11,314	22,873	42,212	33 06	66 12	333
Uranus	4 790	17 00	2,767	21 221	75 01	150 02	3,830	16,129	26,512	27 22	54 44	113
Neptune	4 450	18 00	2,227	22,571	78 45	157 30	6,349	12,890	25,898	37 05	74 09	187

the horizon (now a minimum), and the deviation suffered by the members of the stream which just graze the planet's surface The last column is the same as before. These tables also give, for the respective planets, and on the authority of Professor Littrow, the masses and equatorial radii used in the computations

The table on page 219 shows the immense difference between the results of the perturbations on meteors of the four inner and four outer planets For the first four their attraction can only increase the relative velocity a small amount, for the last four the accelerated velocity is several times the relative velocity. For the first a greater difference is found according to whether the meteor falls from the apex or the anti-apex, for the latter this difference is less and one almost can say that the meteor falls with practically the same velocity. For the major planets the zenith attraction always is very great; meteoric phenomena must show vastly greater complications upon them than upon the earth. These planets can deviate the relative motion of a meteor grazing their surface 150° or more Finally one must note at what immense distances these planets still can appreciably deviate the course of a meteor. In the most favorable case the attraction of Jupiter can change the path of a meteor 4° at a distance only a little less than that of the earth from the sun, while a meteor that went by 608 earth radii away would, under even the most unfavorable circumstances, suffer a deviation of 4°. These effects are only a little less for Saturn, Uranus and Neptune

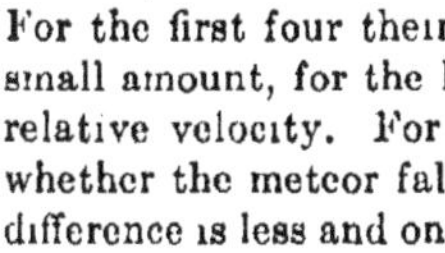

FIG 19

§ 92. When a major planet cuts through a meteor stream, or comes within about 100 radii of the earth from it, the meteors of that part of the stream nearest the planet can be deviated from their orbit so greatly that this part of the stream may be considered as dispersed The result is analogous to that of a storm-wind upon a column of smoke. As the planet moves on the next following, less considerably disturbed, meteors continue to describe the original orbit in space, but the stream shows by a break in its continuity the effects of the meeting, such effects as usually could have been brought about only in the course of a long time If on the contrary the least distance between the planet and the stream is not less than some thousands of the earth's radii, then there is no actual dispersion of a part of the stream, but along a very considerable portion the orbits are more or less strongly deviated and in this portion the stream no longer retains its original regularity. If this should happen several times, finally all parts of the stream would be more or less distant from the original course, so that for the orbits of the individual meteors only a very incomplete parallelism would remain and their bundle be wider and more irregular. Then there would be as many radiant points as meteors, or better we no longer would have a so-called radiant point but only the appearance, well known to observers, called a radiant area in which here and there certain radiation centers for special groups of meteors appear, the units of which had undergone identical, or almost identical, perturbations. Finally it may occur that the original bond between the meteors is no longer provable by observations: such meteors can only be classed as sporadic

§ 93 Not less than the action of these special perturbations, which occur through the nearness of meteor orbits to planetary orbits, are the actions of general and continuous perturbations upon the meteors when within the planetary system or the space immediately surrounding it I will show in the course of this work that a cosmical mass, either a comet or any other body which has a great volume but only a small amount of matter, must be dissipated into a meteor stream in all cases when its own inner cohesion is exceeded by the differential attractions which the bodies of the solar system exert upon its several parts In such cases the comet, say, first will strew matter along its course, and when this path, by virtue of the attraction mentioned, has become elliptical, which it will by the time the dispersion is completed, then will be a closed meteor orbit, and along an arc of it a certain amount of matter strewn Since a small difference of period for the different parts of the dispersed mass is unavoidable, some behind it, some accelerated before it, after a certain time all parts of the meteoric matter will be distributed in a stream according to the lengths of their respective major axes Since these now further come into different positions with respect to the planets of the solar system, they suffer further perturbations in their orbital elements and the orbits which at first were almost identical will deviate more and more from one another. If the first path of all the particles can be considered a single line, then as is easily understood, after a proper time we can regard the new orbits of the single parts jointly as a surface

However, I say further that, if the primitive stream is a linear system (whose transverse dimensions in respect to the dimensions of the orbit are zero or vanishingly small), also after a given time the totality of the stream, in which the orbits of the single parts strew themselves over a surface, will not cease to form a linear system If in fact all parts have from the beginning described a common ellipse, so after a given space of time the variations of the elements of the orbits described by single parts will differ one from another by a single variable, that is to say, about the locus occupied by them in the original orbit at the moment of the beginning of the perturbation Hence the space coordinates of the parts mentioned will differ from one part to another only by this one variable, this however presupposes that all parts lie in a continuous line.

After a certain number of periods the smaller particles will have gained a complete revolution upon the larger If the orbit of the one was identical with that of the other, then the stream would close up into a ring During the interval between the first drawing out of the meteor swarm and the closing of the ring (and this interval can be a very long time) the perturbations bring about more or less considerable differences among the elements,[1] so that in

[1] Only the secular variations would remain the same for all parts However for the long-drawn out orbits the non-secular parts easily can be the more important Also, if it is a question of orbits, which cut those of planets or come very near them, the term secular variation indeed loses its meaning.

general the stream returns back upon itself in the formation of the ring, but without exactly closing and prolongs itself as at the beginning When the smallest particles have gained two complete periods upon the larger, then the stream winds back upon itself in two complete turns and forms two spirals. While this continues, the spirals multiply, and the stream forms a ring-like system. When the orbit of the earth cuts two or more spirals of the system, we have at the same or nearly the same epochs two or more radiants in somewhat different positions. The more the spirals multiply, the fewer the meteors in each, and so much the feebler the corresponding radiants up to the point that one can no longer detect them singly. Instead of a radiant point we have a diffuse radiation, whose meteors come from a more or less extended region of the sky

§ 91 In this way we can perhaps explain the apparent phenomenon of repeated radiation It has been noted for a long time and by many observers that at certain special meteoric epochs neighboring radiants appear in exceptional activity. The August phenomenon is certainly of this nature; in the same class fall the showers which appear from the middle to the end of October, the beginning of December and the end of January consisting of many neighboring radiants, appearing active at the same epoch Double radiants also were cited by Greg in his catalogue Among the radiants observed by Zezioli we can find very obvious examples of this fact, if Nos 1, 6, 12, 13, 18 and 21, which appear from January 6 to 29, are compared among themselves A second group appears to be formed by Nos. 8, 10, 14, 20, 22 and 24, which stretch from January 18 to 31. Equally Nos. 46, 49, 50 and 51, from April 1 to 9, appear to be connected, also the radiants Nos 99, 109, 112, 116, 120 and 126, which appear from July 18 to 31. Finally the remarkable similarity of Nos 132, 137, 142 and 144, which are intimately bound up with No 139, the famous Perseid radiant, is to be noted. A part of these coincidences probably is due to chance, but surely not all. Probably the combined planetary perturbations worked toward this end

(As stated before all the contents of this chapter to this point are a free translation of Chapter VII, in Schiaparelli's *Sternschnuppen*. It has not been considered necessary, however, to include all the footnotes.)

An important addition to the above view of the subject was made by G. Shajn, of Pulkovo, Russia, in an article entitled *The Disturbing Action of the Earth on Meteoric Showers*[2] The perturbations are calculated by the method *Variation of Constants* from Oppolzer's *Bahnbestimmung* II, somewhat simplified. Two showers were chosen as typical, the Perseids and the Bielids, with elongations from the apex of 67° and 158° respectively. For each shower two meteors are

[2] *Monthly Not., R A S.*, **83**, 341, 1923

chosen, the first crossing the earth's orbit in front of us, i.e., before we reached the crossing point; the second crossing behind the earth or after it has passed the crossing point. The meteor in each case was assumed to be 0.00013 from the earth.

	BIFLIDS		PERSEIDS	
	(1)	(2)	(1)	(2)
$\Delta \Omega$	−25 3″	−25 4″	−0 3″	−0 4″
$\Delta \iota$	+1° 35 6′	−1° 27 5′	+1° 6 5′	−34 0′
$\Delta \pi$	+5° 48 4′	−5° 26 3′	−27 6′	+22 6′
$\Delta \varphi$	−10° 1 9′	+9° 20 8′	−4° 9 0′	
$\Delta \mu$	+433 2″	−322 7″	+20 2″	
Δ a	−1 162	+3 087	−7 260	

The following very interesting table shows the changes in the major-axes:

NEAREST DISTANCE TO EARTH OF METEOR	BIELIDS a = 3 541 DISTURBED SEMI-MAJOR AXIS		PERSEIDS a = 24 534 DISTURBED SEMI-MAJOR AXIS	
	(1)	(2)	(1)	(2)
500	1 107	Hyperbola	+7 735	Hyperbola
1000	1 166	Hyperbola	7 790	Hyperbola
1500	1 223	Hyperbola	8 421	Hyperbola
2000	1 278	Hyperbola	8 966	Hyperbola
2500	1 382	Hyperbola	9 972	Hyperbola
3000	1 478	+68 027	11 148	Hyperbola

To find the resultant change upon the radiant due to a single passage he calculates:

	BIELIDS		PERSEIDS	
Δ l	+3° 31′	−12° 30′	−1° 0′	x
Δ b	+4 34	−2 40	0 12	x

From this he concludes that for a shower with considerable elongation a few revolutions are sufficient for the formation of a considerable area of radiation. He states that as the Perseids are very old, at their moderate elongation their dispersion is large. For the Orionids and Lyrids, being near the apex, it should be small. Following this reasoning he shows one could expect to find few showers connected with the short period comets, and that those we do find soon will be scattered, as indeed has happened for the Bielids. He feels that his work, showing that the major axes change so greatly, completes and generalizes the reasonings of Schiaparelli on the spreading of an initial cluster due to the sun's action He also states that for showers with large elongations we should meet meteors at both nodes, due to the scattering action of the earth, and believes that such eventually will be discovered in the southern hemisphere. His summary is:

1 Many details of the complicated phenomenon of meteoric showers must be considered mainly as a consequence of the disturbing action of the earth.
2 The perturbations depend in considerable degree on the conditions of the passage of the meteor through the sphere of action of the earth.
3 The general consequences are (*a*) a formation of an area of radiation, (*b*) a peculiar form of this area; (*c*) the stability of the showers and the peculiar distribution of the rich showers, (*d*) the feeble and dispersed radiation in the vicinity of the anti-apex, (*e*) the spreading of the meteors along the orbit, (*f*) the polarity of the radiation, (*g*) the abundance of the meteoric showers and the number of comets, (*h*) an increase of the density of the meteors with approach to the sun as a consequence of the secular disturbing action of the earth.

This article furnishes a welcome addition to our knowledge of all the subjects mentioned, and the writer heartily agrees that in general its conclusions appear sound. Nevertheless it is believed that the total possible effect, at least for cases like the Perseids, has been over-estimated. For instance taking the values given for the perturbed semi-major axes of the Perseids at distances of 1000, 2000, and 3000 km., respectively, we find that the changes from first to second is 1.176, and from second to third 1.186, in other words about 1.18 and nearly constant. If this rate of increase holds (which is not true but will serve for a first rough approximation) at a distance of about 15,000 km. the action would almost cease. Adding in the radius of the earth and throwing in something extra due to the approximation, let us say that 30,000 km. from the earth's center is a fair estimate. Then the earth can perturb all the meteors in a cylinder of space with a cross section of π (30,000 km.)^2 and a length equal to the distance it goes while passing through the stream. But this in turn is the *least* cross section of the stream, if the cross section is assumed to be circular, multiplied by the secant of the angle of inclination. Or the cross section of the stream has as a minimum diameter the distance travelled by the earth in going through it multiplied by cos ι. This distance is for the Perseids about 80 million km., if we assume it takes a month to go through them. When multiplied by cos 116° this drops to about 72×10^6 km. The minimum cross section would be π $(36 \times 10^6 \text{ km})^2$ in area. The period corresponding to a = 24.534 is P = 121 years approximately. Finally if we assume that the earth cuts the stream along a diameter of the cross section, and the latter is not moving, then the earth perturbs only one meteor in every 1200 of that particular cross section of the stream. When we remember the number of such cross sections that are to be found in a stream with a = 38×10^8 km. long, the cross sections being only 60,000 km. thick, we see that hundreds of million of years would be necessary for the earth alone to produce such a distribution as found by observation if the distribution is caused entirely according to this theory.

It must be clearly understood that the figures used in this discussion are rough, and the assumptions the simplest, but while modifications would indeed change the numerical results immensely, the general argument would not be appreciably affected. As we have no right to assume that the Perseids are of any such age as indicated above, this theory alone can explain only a part of their present distribution, but by no means the major part In such cases as the Bielids the theory of Shajn has a far wider application, and doubtless is fully adequate to explain a large part of the actual distribution. Its importance as a real advance is therefore again emphasized.

CHAPTER XIX

Influence of Meteors upon the Earth

No matter whether the total mass of the meteors which meet the earth in a given time be great or small, there are certain effects which are obliged to follow The most obvious is that the mass of the earth is constantly increasing. By well known laws this increases the attraction between the earth and the sun, which in turn decreases the length of the year. Again the continual meeting of the earth with meteors has the same effect as if we were moving in a resisting medium and, paradoxically, makes the motion faster This is because, while the medium does tend to slow up the earth, the earth as a consequence drops slightly nearer the sun, which in turn requires it to move faster. Hence the paradox that a resisting medium causes an acceleration. As stated, the year is made shorter by both of these causes.

There is an effect also upon the length of the day. This is due to the fact that the meteors pile up upon the surface of the earth only. The earth has a certain moment of momentum, which cannot be increased, about its axis. As more material is added, while the energy of rotation remains the same, obviously the rotation cannot be so rapid. Young states[1] that the combined effect of all three causes does not amount to 0.001 of a second in a million years. It should be added that probably a greater mass fell per year in by-gone ages, and also that even yet we are far from having any exact figures upon the total yearly mass that is still falling.

Another obvious effect is the heating of the atmosphere due to the energy liberated within it by each entering meteor. It is scarcely necessary to add that, using any reasonable mass for the calculation, the results will be very small. Young on the assumption of a daily fall of 100 tons of meteoric matter calculated that the sun would furnish as much heat in 0.1 of a second as this mass of meteors would in a year. Hence it is hopeless to expect the detection of such an amount by direct measurement

[1] *General Astronomy*, 475.

Another effect upon the earth's orbit is pointed out by Bosler.[2] He says: "While our distance to the sun is diminishing, the inclination of our orbit to the invariable plane of maximum area diminishes also, provided indeed that, in the definition of this plane, meteors are taken into account." Also in speaking of other causes he points out that the earth's orbit is really a spiral, due to its constantly, if slowly, diminishing distance from the sun. In the same article a very interesting discussion of the effect of meteors upon the eccentricity as well as upon the inclination of the earth's orbit is undertaken. By analogy this is extended to the orbits of other planets. His argument is somewhat complicated because of the belief that most meteors move in direct orbits, which elsewhere in this book we have attempted to show was unproved, even if not untrue. Had he assumed an equal or even larger number of retrograde orbits, which we believe is the correct assumption, the argument would have been much strengthened But despite this handicap he develops a theory that with time the eccentricity of the earth's orbit also diminishes. And in general that perhaps we have here a true reason for the most striking characteristics of the solar system: orbital motions direct and nearly circular, and almost in the same plane. The article is particularly interesting in any study of the evolution of the solar system and the possible rôle of meteors therein.

[2] *Bul. Soc. Astr. de France*, **33**, 241, 1919.

CHAPTER XX

DISCUSSION OF THE FORMATION OF THE BIELID AND PERSEID STREAMS

In previous chapters a great deal has been said about both the Bielids and the Perseids, yet it is desirable to examine further how the streams might have acquired their present constitution, in the light of the theories already given. The main facts as to Biela's Comet and its orbit have been set forth at length in Chapter VII, so here we need only add that it comes within 0.4 of Jupiter's orbit, and that about every 60 years it and that planet are near each other. Such passages produce great changes in the nodes, which indeed begin when comet and planet are at considerably greater distances than that mentioned. As the comet barely misses the earth's orbit, and has a direct motion with an inclination of 13° only, the earth perturbs it tremendously, and we cannot expect groups moving in such an orbit long to preserve their cohesion. Hence for any given part of the stream, if it once furnished a splendid shower, that exact part would be so dispersed by its passage that we hardly could expect an equally good shower from it in the future. For instance, individual meteors could have their periods changed between limits of 1.7 and 390 years Then for meteor showers to come from groups moving in the comet's orbit, naturally they must have nearly the same period as the comet, and we must consider that those particular groups have not yet been within our sphere of influence, else they would have been at least partly dispersed.

From these points in view the Bielids of 1798 and 1838 might have come from a fragment detached in 1772, which had gained 4 months on the comet itself by 1798 and 7 months by 1838.[1] For the meteors of 1741, 1830 and 1847, which groups are very far removed from the comet,[2] the older theory of perturbations rather than of detachments must still be employed. In 1841 Jupiter again greatly perturbed the comet. But Bielids seen in 1850, 1866 and 1867 were all from groups distant from the comet. Those seen in 1852 were observed 3 months

[1] Schulhof, *Les Étoiles Filantes*, **57**, 1894.

[2] *Bul. Astr.*, **11**, 126, 1894

after the comet had passed. Schulhof further concludes that none of the great showers of 1872, 1885 or 1892 could have come from the débris of the two nuclei seen in 1846 and 1852, and hence that it is necessary to admit the existence of two unknown fragments, supposing that the swarms of 1872 and 1892 are identical. Slow disintegration could not explain the retardation of 11 weeks from 1852 to 1872 That the disintegration of the two nuclei took place without much change in their periods also adds to the difficulty. For instance, the earth in 1865 passed the node only 4 weeks ahead of the comets, but no shower was observed. And if the comets had by then been greatly dispersed meteors would certainly have been seen.

We stated that Biela's Comet was never seen after 1852. It is but fair to state, however, that several English observers in November, 1865, reported faint objects, in at least one case in motion, which they believed to be cometary and were near the ephemeris positions of the comet, which was being widely searched for at the time [3] Bad weather and other circumstances in all cases prevented their confirming the observations in the usual way. Pogson's observation at Madras in 1872 has already been given. It is unfortunate that these observations of 1865 did not agree well among themselves, and the roughness with which they were made permitted little weight to be attached to them. Yet they deserve mention.

Schulhof adds that the meteors of November 24, 1872, seen three days before the main shower, must have come from a third little fragment whose node was more changed by Jupiter in 1841 than the node of those which formed the great shower of November 27. On the contrary it can be admitted that the few Bielids observed during the years between 1872, 1885 and 1892 were detached from the two or three principal swarms by the earth The two known nuclei would have been nearer to Jupiter in 1901 than in 1841, and therefore from that time great modifications certainly could occur in any future node passages of débris from them. Actually the fewest Bielids have been seen since 1899.

H A. Newton, 1894,[4] presented a hypothesis about the Bielids which if true could be extended to all streams. He starts with the theory that the great showers of 1872 and 1885 came from bodies

[3] *Monthly Not*, *R A S*, **26**, 241 and 271, 1865 6.
[4] *Am. Jour. Sci* (3), **47**, 152, 1894.

detached from the comet after 1840, otherwise the radiant would be the same as for the meteors of 1838. As it was, the orbits of the meteor streams from the new radiants fit the new orbit of the comet. But in 1872 the comet was 2×10^8 miles away from the node when the great shower appeared, in 1885 it was 3×10^8 miles ahead of the node. So that some of the particles leaving the comet between 1840 and 1870 had fallen behind and some gained. He further stated that whatever force acted must have done so only in the plane of the orbit, otherwise the groups would have been scattered and no great shower would have been possible, as this could only come from a compact group. This force must be a repulsive force, and only the sun could be its seat for it alone is always in the orbit plane. Analogy to the forces active in forming the tail was then brought forward. Yet according to this some must go forward, some backward from the comet.

He suggested, as a possible explanation, that each separate unit has a load of electricity, by which they have a permanent repulsion or attraction "sufficient to change the orbit altogether, not in kind, but in a steady change, throwing them into a new orbit with a new period, and then scattering them." As the load of electricity, or whatever force it is, presumably is not lost, the action continues. Hence even after the total destruction of the original comet, new swarms may be formed.

It is very difficult to pass an opinion upon such a hypothetical force under such little-known conditions. Yet it is true that repulsive forces are at work within comets. Not only the tail matter but also other material, as that in the jets, for instance, is ejected, hence Newton certainly had at least a partial foundation for his theory. The most difficult thing about it to accept is the continuous action demanded, action that in a very few years would carry great groups of particles which would follow exactly the same orbit, 2×10^8 miles from the parent body, and yet that this action would not have enough power to totally scatter the groups themselves. The fact that the particles still follow an orbit practically the same as that of the comet, with so few years given for such great effects, seems to be an almost unsuperable obstacle to accepting this hypothesis. Any theory which would give us more time for the dissemination of the material along the orbit would, it appears, have the advantage for the reason given.

Brédikhine in his discussion took the case of Biela's Comet, and, assuming for r = 1, $\gamma = 0.1$ and $\gamma = 0.2$ respectively, he calculated the new periods of the emitted masses, for various values of J. In this way he showed that it would be very easy to get accelerations of several months and retardations equally large. He then definitely calculates that on the hypothesis of nuclear emissions the three groups of 1877, 1885, and 1892 could have been produced about 1846 by three masses detached with velocities of 0.292, 0.342 and 0.279 km./sec., which would give periods of 6 718, 6 645 and 6 695 years respectively. Brédikhine finds strong support for the details of his method because the radiant area of the Bielids in 1885 was an oval, not a circle, with the major axis running north and south. As this area had about the expected size and its axes coincided with those calculated on his theory, the accord between the latter and observation was good

Schulhof remarked that if we accept this theory in general we should expect no more great showers of Bielids. Because it would be inadmissible to attribute as strong emissions of matter to the fragments of the comet, as could be legitimately done for the large comet itself. He concluded that, in his opinion, the hypothesis of slow disintegration is sufficient to explain the facts, at least if we give several centuries of time. He added that there are other comets, whose elements are not well known, which greatly resemble those of Biela's Comet, and this fact might well go to prove that disintegration of the original body had long been in progress.

Turning now to the Perseids we reach the most difficult case. This is true because if we accept, as we must, that a succession of radiants may be found moving in the same direction and including those on the maximum dates of the Perseids, then the stream must have an immense width. Denning gives six weeks as the certain duration of the shower. The earth meantime has travelled about 12 per cent of its whole orbit, a distance of some 70 million miles. The perihelion distance of the group as calculated by Schiaparelli is 0.96, which is passed July 23, when we have already entered the stream. The corresponding width is almost inconceivable, if we assume that Tuttle's Comet is indeed the generator of the stream.

It is here necessary to say, however, that the long duration of the Perseids, i.e., supposing all the various radiants included in the ephemeris to be Perseid, is not universally admitted. The person who brought out a possibly different interpretation was Brédikhine, who concluded from his investigation that the radiants observed in July, and those probably after August 19, really belonged to other streams, or were partly chance accordances. He reasons that an hourly frequency of only one or two meteors, as given by Denning up to July 28 and after August 17, for a region which at that time of the year contains many radiants, is too low to make it certain that these meteors really all belong to any one stream.

The writer had never, until writing this, seriously questioned the correctness of Denning's whole ephemeris, but having read the above criticism the data secured by the A. M. S. were examined, which include all his own observations. It is found that many radiants from the work of others, and some of his own, confirm the ephemeris from July 21 on, but further, since so many other radiants are found in the lists, whose coordinates do not differ greatly from the ephemeris positions, that it is partly a matter of selection In other words the number of contemporaneous radiants in the general region α = 10° to 50°, δ = +40° to +60° which do not conform are rather numerous. Nevertheless, the work of the A. M. S. may, on the whole, be considered to substantiate the existence of radiants in the positions assigned by Denning to the Perseid radiants from July 28 on to August 18. The data are not extensive enough to give a positive answer either way from July 21 to July 27, and after August 18 a negative answer can be given.

Brédikhine's idea of a radiant, already referred to, which gave it a great size if we judge from the manner in which he reduced the Russian observations of the Perseids in 1893 and 1894, doubtless colored his theories somewhat. Yet he very sagaciously pointed out the results of wrong conclusions of others based upon their observations. This apparent inconsistency was due to his taking the radiants published in Denning's *General Catalogue*, along with the excessive areas often found by the latter for stationary groups, as actually existing as published While he expressed vigorous doubts about much of the data, still he employed the radiants in his calculations.

Returning to his treatment of the Perseids on the emission theory, he assumed the area of radiation as 10° in diameter for August 19 and obtained a value of j = 0.17. He thought that August 14 or 15 was about as late as we should see true Perseids. He felt that the position then given for August 19 by Denning, which was $\alpha = 65°$, $\delta = +57°$,[5] too far removed from what his results gave and probably was due to too few observations. But for July 25 the value $j = 0.4$ is required, which he considered out of all bounds. But in adopting $j = 0.2$, all periods from 3 to 112 years will be found for J varying between $+45°$ and $-9°$. As the different swarms result from different emissions, some can approach the exterior planets and undergo strong perturbations which will much augment the extent of the faisceau or bundle of orbits where the earth crosses it.

Schulhof remarked that such a conical faisceau of orbits, as required by the original hypothesis, without the help of planetary perturbations, can, for admissible values of j, have a cross section of only a few degrees. This could be crossed by the earth in a few days, but with perturbations its size could be vastly increased. Also that the hypothesis presents the advantage that the diversity of periods makes meetings with major planets more frequent But the disadvantage that such clusters coming from nuclear emissions would have a feeble cohesion and would be scattered as a consequence of such perturbations. Hence he believes the emission theory a powerful adjunct only of the perturbation theory, i e , the truth is a mixture of the two.

Schulhof in his own discussions noted that Tuttle's Comet has $\iota = 114°$ (therefore retrograde motion), that it can approach the

[5] The later position given by Denning, $\alpha = 57.4°$, $\delta = +58.5°$, would doubtless fall in with the theory well enough It is necessary here to call attention to a great discrepancy in the three lists of Denning's observations which appeared in *Monthly Notices*, *R A S* , **45**, 97, 1884; **50**, 438, 1890, and **62**, 163, 1901 respectively. From his own observations only he derived.

DATE	1884 LIST		1890 LIST		1901 LIST		NOTES
	α	δ	α	δ	α	δ	
August 14	54°	+57°	53°	+57°	51°	+57°	In 1901 list the averages are taken for each date
15	56	57	—	—	53	56	
16	59	57	60	59	53	58	
17	62	57	—	—	54	60	
18	65	57	—	—	55	59	
19	68	57	—	—	57	58½	

In other words all of these positions from August 14–19 of the 1884 list, and one of the two in the 1890 list were discarded in the 1901 list On what grounds this was done is questionable, because positions as early as of 1877 do appear in the 1901 list.

earth by 0.005, but Saturn only by 0.75. He proved that a meteor moving in a orbit with the elements of this comet, grazing our atmosphere, will be deviated only 1.6°. But to be so perturbed by Saturn as to cause the changes in inclination of the comet's orbit great enough to give such a width to the stream, it would have to come to within 0.01 of the planet and be deviated 10°. The deviation due to Saturn could be 74° at distance of 1.5 radii of planet. To have all possible values between 10° and 74°, the inclination would previously have to be very small. He then undertakes a mathematical discussion of the case which was based upon a method given by him for planetary perturbations But this led him to such results that it would have been necessary for the comet and its numerous débris to have moved in retrograde orbits almost in the ecliptic, with a relatively short period. From the moment the perturbations could have augmented the inclination, the node of the new orbit, which must be very near perigee, would become more stable. But it is most difficult to admit that, after many modifications of the primitive orbits, the resultant orbits could be so alike as those of the Perseids seem to be. He further remarked that, if the earth and another planet are both passed by the stream a necessary result is that the parameter p of the stream's orbit equals $\frac{2\,R\,r}{R + r}$, where R and r are the radii vectores of the planets in question.

Schulhof continued that then the parameters of different groups of Perseids must be equal, and also the other elements, if their inclinations are equal. But for orbits calculated by Kleiber for the radiants given by Denning, the values of ι are very different, with q almost the same. Then if his (Schulhof's) hypothesis were true, these streams might all have had a common origin with the Perseids—but he considered his hypothesis rather impossible. The great difficulty is that it required the stream to approach to within 0.01 of Saturn; actually the distance was 0.75. It would then be necessary to suppose that the stream during centuries could have undergone slow perturbations by Jupiter and Saturn which, added to modifications of the orbit of this last planet, might have made its point of closest approach to Saturn continually move further out from the earth's orbit. He concluded his article by some computations based on Brédikhine's hypothesis of emissions only and found such abnormal values for j that he considered them impossibly large.

Antedating this last research by two years Kleiber in 1892[6] published a most important paper, mentioned elsewhere (p. 170), in which, among other things, he considered the daily motion of radiants belonging to a meteoric ring. He stated there are two views: (1) the ring (or part of the ring) forms a definite part of the solar system; (2) the meteor stream may be considered as the result of the disintegration of an agglomeration of independent cosmical particles from outer space. In the first case elements ι, $\pi - \Omega$, (and ϕ if for an ellipse) must be constant. In the second case only the heliocentric longitude of perihelion is constant, the other elements vary. He considers the first hypothesis true for the Perseids and derives the following rule: the latitude of the radiant must remain constant, its longitude increases proportionately to the time. Or the radiant moves along a circle parallel to the ecliptic.

He applied this rule to Denning's positions given in his 1890 list[7] and found that of 49 radiants there called Perseids, 46 lie within 2° of the point which he calculated as the cometary radiant, a result that he justly considered very good. We must note that Kleiber here had the advantage of Denning's second list from which the latter had omitted the discordant radiants, i.e., those among August 14 to 19, which had caused such trouble to Brédikhine in his 1888 work on the same subject. Kleiber therefore felt that Denning's position, as given in 1890 well confirmed his theory. Had Kleiber lived to use the 1901 list, better accord still would have been gotten since in that list Denning again omitted the positions given in the 1890 list for July 8 and August 14, which were two of the three radiants which gave large residuals as shown in Kleiber's article published in 1892.

It is regretable that the case of the formation of the Perseids has not been solved in a complete and satisfactory manner. Yet it is far better to state the fact frankly than claim too much for our present meagre knowledge. It is certain that parts of the truth have been brought out, and with advances to be expected in the next century doubtless we then will be much nearer the whole solution. The labors of the various eminent men mentioned have laid a solid foundation for advance in the right direction.

The Lyrids, $\iota = 80°$ approximately, do not last more than 3 or 4 days, and usually are very infrequent except on one day. Hence the

[6] *Monthly Not*, *R A S*, **52**, 341, 1892.
[7] *Monthly Not.*, *R.A S*, **50**, 410, 1890.

maximum extent is only 4°. The parent comet, 1861 I, has a period of 408 years and comes within 0.3 of Saturn. This planet therefore has modified the period of different particles The earth doubtless has contributed considerably to the scattering action, as the motion is direct and the comet approaches our orbit to within 0.002.

Temple's Comet, the one which moves in the same orbit as the Leonids, has $\iota = 162°$ and P = 33 years It approaches the orbits of Saturn and Uranus by 0.45 and 0.4, respectively, and that of the earth to within 0.007. Its particles are concentrated essentially within a 3° arc of our own orbit, but a few stragglers stretch this distance to perhaps 7°. Due to the circumstances mentioned more scattering might be expected than actually is found, if our theories are approximately correct.

Halley's Comet and the η Aquarids move in an orbit with $\iota = 162°$ and an average P = 77 years, at least for the comet. We know that perturbations are sufficient to change the period of the comet itself from 74 to 79 years, hence we need no indirect proof here that the bundle of orbits in which dispersed particles move must have a considerable width, and that the individual particles have periods differing by several years among themselves. We are therefore not surprised to find the earth taking at least 11 days to pass through this meteor stream. The question already has been somewhat discussed in the chapter dealing with these meteors The comet's orbit has nearest approach to that of certain planets as follows: Earth 0 05, Mars 0.05, and Jupiter 0 8 The various distances of approach to planets' orbits, quoted for these three cases, are taken from Schulhof's table appearing in *Bulletin Astronomique,* **8**, 291, 1891.

Nearly all the data on these η Aquarid meteors have been obtained so recently that no other writer of the longer memoirs quoted had the opportunity to discuss them. All these memoirs appeared before the articles by the writer and Hoffmeister put the connection of these meteors with Halley's Comet beyond doubt. The excellent records we have of this comet for so many centuries, the proof that its period can change nearly two years on each side of the average, and the further remark by Crowell and Cromellin that gravitational causes are quite sufficient to explain these great changes, form, in the writer's opinion, a clearer chain of evidence of connection between a comet and a stream of meteors at some distance from its orbit (0.05 as a minimum) than any of the classical cases. If the comet's period could thus change, we know absolutely that the periods of the bodies, which left it in time past, can have periods now differing by at least four years, with corresponding differences of major axes and other elements, due to the particular perturbations each particle underwent. But we still find these particles scattered, if sparsely, out to a distance of about 0.10 from the comet's orbit, forming a bundle of orbits, in exactly the same manner that others have proved for the older cases. Only here the radiants were all carefully observed, we do not get an unbelievable width of the stream from the observations, and accurate orbits were at once calculated for the different radiants.

This case therefore probably is more easy to investigate, and we should not hesitate to accept the verdict of the two eminent English astronomers mentioned, that gravitational causes alone could bring about the changes in the comet's orbit and, inferentially, in those of the particles. For this stream, at least, we are without the difficulties encountered in explaining for the Perseids. We find what we might reasonably expect. The only hypothesis needed is one to explain how a particle, originally part of the comet itself, became separated. Once this separation is effected, the rest is clear. But here we have two excellent causes, long since given by others, the scattering action of the sun and planets upon such an aggregation of particles as we assume a comet's nucleus to consist of, and nuclear emissions, which we actually see taking place in some comets in the form of jets. Halley's Comet and the η Aquarids may well stand henceforth as the simplest possible cases of the gradual formation of a meteor stream from a great comet.

CHAPTER XXI

METEORITES

Meteorites generally are divided into three major classes, aerolites, siderolites and siderites, stony, stony-iron, and iron, respectively, in composition. While many more stone meteorites than any other kind are seen to fall, yet museums contain more specimens of the iron ones. This is due partly to the fact that a stone meteorite may look too much like a terrestrial rock even to attract attention, while an iron mass found in a region where, or position in which, it was not to be expected generally would excite curiosity or comment, even by untrained persons. Those who have made a special study of the structure and chemical composition of such bodies usually are able quickly to decide whether a specimen submitted to them is of celestial or terrestrial origin. But in certain cases authorities have differed, hence the origin of a few masses found in collections still is doubtful. The proportion of these cases is, however, fortunately not very large.

Even when the falls were seen and the pieces quickly picked up by eye-witnesses, the greatest variety is found in size, number and composition. In many cases only a single mass is known to reach the ground, but in others as, for instance, that at Holbrook, Arizona, July 19, 1912, a careful count and estimate gave 14,000 fragments, while the famous fall at Pultusk, Poland, June 30, 1868, consisted of perhaps 100,000 fragments. In all these cases it must be clearly understood that most of the fragments were very small. The total weight of the Holbrook fragments was only 218 kilos. But on other occasions large individual masses have fallen as at Knyahinya, Hungary, June 9, 1866, when one of the masses weighed 293 kilos. The total weight of this fall was about 476 kilos. The largest iron mass, actually seen to fall, is that from Cabin Creek, Arkansas, weighing nearly 47 kilos.

If we turn to masses of iron, found under such circumstances and of such structure that their meteoric origin is certain or very probable, vastly larger individuals occur. The largest of all belongs to a group of three found on the shores of Melville Bay, Greenland. They were brought to America by Peary, later of North Pole fame, in 1895 and 1897. They are now in the Natural History Museum, New York City The largest weights 36½ tons, with approximate dimensions of 10.9 × 6.8 × 5.2 feet. Their fall was unrecorded but the natives certainly had known of their existence many years before 1818 when Captain John Ross first inquired about them.

The other great ones are, in order, Bacubirito, Mexico, 27 tons; Chupederos, Mexico (2 pieces), 26 tons; Willamette, Oregon, 16+ tons, El Morito, Mexico, 11 tons, Bendago, Brazil, 5 tons; Cranbourne, Australia, 4 tons (largest mass of several), and about eight others weighing a ton or more, all found in North America except one —that of Magura, Hungary. Many of the Mexican meteorites have been known for over 300 years and some of them certainly were considered holy by the Aztecs and preserved in their temples. In the United States splendid collections of meteorites are at the Field Museum, Chicago, The Museum of Natural History, New York; and the National Museum, Washington. Yale University has a very large collection, while Harvard University, Adelbert College, and Amherst College have good collections. Smaller collections are in Milwaukee, at the University of Minnesota, in Philadelphia and St. Louis. The Mexican National Museum contains some of the remarkable meteorites just mentioned. In Europe there are superb collections in Berlin, London, Paris and Vienna. Many other European cities contain fine collections of different sizes, while in Japan and various portions of the British Empire small collections are now being formed.

The finding of meteorites depends upon so many circumstances, besides the fact of fall, that an explanation of why they have been found in certain localities and not others becomes difficult. Farrington[1] states that of 634 known meteorites, 256 have been found in Europe and 177 in the United States, or about 68 per cent in only one-eighth of the land area of the earth The most obvious reason is that here we have the most educated populations and much of the area fairly densely settled. He further states that of the 256 from the western hemisphere 182 were iron and 74 stone, while from the eastern hemisphere of 378, 299 were stone and only 79 iron. These different proportions are very difficult to explain. Again it is obvious that on a desert surface meteorites would be very easy to find, as in

[1] *Meteorites*, 1915.

Arizona for instance near Coon Butte, for they would last longer in the dry air and there would be less vegitation to cover them. But a very large number of meteorites have been discovered in the Southern Appalachian region, in Virginia, North and South Carolina, Georgia, Alabama, Kentucky and Tennessee. This region has a fairly moist climate, many of the mountains and valleys are densely covered with vegetation, and the population in much of this region is not large. Meteorites here certainly would decay or disintegrate quite rapidly, nevertheless numbers have been found. On the contrary in some of the central states, which are quite level and with denser populations, few if any meteorites have been known to fall. As for the Rocky Mountain region, the population is so small that we hardly can compare it with either of the regions just mentioned. While such a distribution of falls most probably is due to mere chance, yet it is remarkable enough to be worth mentioning.

From about 1800 on there was some chance for every fall seen by civilized men to be recorded. We have thus divided the falls into ten year intervals, beginning with 1801 and including 1920.

1801–10	16	1831–40	20	1861–70	49	1890–1900	36
1811–20	18	1841–50	29	1871–80	46	1901–1910	44
1821–30	24	1851–60	37	1881–90	40	1911–1920	36

In 1868 no less than 11 were reported, while 1865, 1877, 1886, and 1910, had 7 each—the next largest number. It is striking that the maximum for the 10-year intervals fell between 1861–70. As the educated population of the earth has vastly increased since that time, this cannot wholly be due to chance. Up to 1923 about 850 falls and finds had been collected and were available for study. Of the siderites only 14 were actually seen to fall.

The angular size of meteorites and fireballs frequently is so great, as reported by competent observers, that we would derive linear dimensions of not less than one or two kilometers for some of them. It needs no discussion to establish the fact that if bodies of such size entered our atmosphere they would reach the earth with only their surfaces seared to a slight depth, as is actually found for most meteorites. Irradiation indeed plays some part; for instance it is hard to realize that the planet Venus has no angular diameter visible to the eye as such, yet its maximum is only 67″, a quantity too small for the eye to see except as a point. But this is not sufficient to explain the great diameters observed.

As was pointed out in the theories given concerning the effect of the atmosphere upon a meteor and the formation of its train it was seen that the meteor carries with it a cap of intensely heated and glowing gas, part of which must constantly escape. All this would tend to increase its apparent diameter. But the writer frankly confesses his unbelief in the adequacy of any of these theories to explain the phenomenon, *if* the meteorite consists of only one mass. The above reasons are indeed correct but seem hardly sufficient. This brings up another phase of the problem which can be stated in a question—does a fireball or meteorite of very large angular diameter consist of only one member? In many cases a positively negative answer can be given. Farrington in discussing this question[2] gives four good reasons for, and an equal number against, the average meteorite coming within the sphere of the earth's attraction as a single body. In favor of unity he gives angularity of individuals, uniform composition, appearance as a single body, and narrow distribution of components. He himself partly answers the third by quoting the case of the Rochester meteorite. His reasons in favor of many members are: complete encrusting of most units, small number of showers, regular form of area of distribution, and difficulty of breaking up an iron mass.

In explaining cases of either meteorites or fireballs of very large angular diameter, which postulates large linear dimensions, there seem two alternatives: either there are many units which are really at distances up to perhaps 100 meters from one another, or some other action than the mere turning of kinetic energy of motion into light and heat is at work and causes the atmosphere for several hundred meters around the mass to glow. Or both may combine in some cases, for if the second is true the glow would appear whether one or many bodies were present

Let us suppose that many units are present and see whether Farrington's four reasons for the assumption of one body can be partly removed. There is reason to believe that a comet's nucleus is made up of just such angular fragments and that meteorites also are fragments of larger bodies and that in such fragmentation units large

[2] *Meteorites*, 1915.

and small could leave the broken body in groups. If these statements may be accepted there is no difficulty in explaining the second point, uniform composition. As for its appearance as a single body, if there were a large number of units in a group, each with its glowing cap of gas and heated atmosphere around it, then the whole, at a distance surely would appear as one glowing body, which usually is the case with the bodies under discussion The last point "narrow distribution" is interpreted to mean distribution over a small area. This is to be examined in the light of data on that point. For instance the Knyahinia fall was scattered over an area of 9 × 3 miles, the Khairpur over 16 × 3 miles, and several others over an area of at least 6 to 20 square miles. If these represent the cross sections, even very oblique ones, of the groups, our hypothesis is more than satisfied, i.e , that they came as groups of separate units.

It will be of interest to give here a brief description of the most remarkable group of meteorites ever observed as such. It was first seen at 9:05 p.m. on February 9, 1913. It began at least as far west as Saskatchewan Province in Canada and was last seen by ships at sea beyond Bermuda. Full accounts were published in the *Journal R. A. S. of Canada*, **7**, 1913, by C. A. Chant and others, and more recently the whole evidence has been reviewed by W. H Pickering in *Popular Astronomy*, **30** and **31**, 1922–3 It seems to have taken 7 minutes to travel across Canada The latter writer believes the group struck the sea some seven hundred miles beyond the steamer from which they were last observed, making a total path of 6000 miles. There seems to have been at least 10 groups, with 20 to 40 members per group. In Canada it is said to have taken 3.3 minutes to pass a given place. The geocentric speed over Canada was between 5 and 10 miles per second. One observer states that only the first great meteor of the procession burst, the others were already in groups. They left long trails. Their passage was accompanied by sounds like distant thunder, and in some localities the earth vibrated. The outstanding feature was the perfect formation kept by the members of each group, and how group followed group in exactly the same path. It is most unfortunate that despite the great number of observations in Canada and at sea, only one or two people in the United States seem to have seen this wonderful phenomenon, though it passed over the most populous part of our country.

Returning from this digression, we may agree with the reasons of Farrington favorable to the group hypothesis except to say that the second, i.e., the small number of showers, is not clearly understood as to meaning, hence cannot be discussed. However, fairness demands that certain partially adverse statements be made. In some cases when fragments of a meteorite have been picked up, it was possible to actually piece together the original, the faces where cleavage took place being clean, while the original surfaces had the usual coating due to the melting incident to its passage through our atmosphere. Therefore during this part of the flight, when we have reason to believe they were brilliant, they consisted of one mass We cannot carry the argument further unless we have positive testimony as to the apparent brightness of the individual meteorites which behaved thus. This cannot be furnished here, nor is it certain that it is available in most cases.

Returning now to the three classes of meteorites already enumerated, the aerolites consist essentially of silicate minerals with minor amounts of metallic alloys and sulphides; the siderolites of a network of metals with the spaces between filled with silicate minerals; the siderites mostly of iron and nickle, or their alloys, along with iron phosphides and sulphides. Taking meteors as a whole the following elements are found abundantly. Aluminium, Calcium, Carbon, Iron, Magnesium, Nickle, Oxygen, Phosphorus, Silicon and Sulphur. Sixteen other elements including the two elements Hydrogen and Nitrogen have been found, and probably Argon and Helium besides. Seven other elements have been reported but doubt about their actual presence has been expressed. The fact that no new element has been found in them gives strong confirmation to the belief in the essential chemical unity of the universe.

If we turn however to minerals which occur in meteorites, while we find many familiar to us such as olivene, pyrites, magnetite, etc., at the same time very numerous new minerals are found, in one list no less than fourteen being given. Indeed nearly all the elements occur in combined states, yet a few have been found also in their elementary condition. It is curious that quartz, one of the most common terrestrial minerals rarely is met with in meteorites. In general, analysis proves that meteoric stones belong to a class of rocks low in silicic acid but high in the basic constituents, iron and magnesia. The metallic meteorites consist essentially of alloys of iron, nickle and cobalt, and frequently also the phosphide schreibersite and the sulphide troilite. Rarer elements also are found in small quantities in this class.

Meteorites usually are found coated with a black crust or varnish such as would be caused by strong heating. Its thinness shows that the heat due to its flight through our atmosphere had a chance to penetrate only a very small distance during the brief time available. Indeed in the case of stones most of this surface must have actually been driven off from the main mass during its flight and helped to make up the train or trail of sparks sometimes seen. While usually black, this crust sometimes is dull or grayish. This crust is indeed a more or less perfect glass but rarely is more than a few millimeters in thickness. In some stony meteorites are found small, threadlike veins evidently due to a fracturing of the stone long before The dark filling material is of undetermined origin. If the surface of an iron meteorite is polished and exposed to the action of acids a very beautiful etched structure is shown. The figures are known as "Widmanstatten" figures and usually though not always may be found. Their presence was once thought a proof of meteoric origin, but at present this is not considered sufficient in a given case.

The surfaces of most meteorites show cavities which are known as pittings. They evidently are caused by the melting of the most easily fused materials, which then leaves these hollow cavities. There is some difference between those on the front and those on the rear of a meteorite, those in front being small and deep, those in the rear being broader and shallower.

The shapes of meteorites vary greatly, perhaps the most usual being a sort of cone or pear-shaped figure. In such cases it is presumed that the apex represents the front of the mass, being in fact formed by the fusion of the edges if the body entered our atmosphere as an irregular block. In at least one case, the Tucson meteorite in the United States National Museum, the form is that of a complete ring.

The elements in meteorites are the same as in terrestrial rocks yet frequently the form of combination is very different and of a nature to indicate that they formed under conditions not now existing upon the earth, particularly with reference to the presence of moisture and free oxygen It is impossible to believe that phosphorus, the metallic nickle-iron and the unstable sulphides can have formed and remained unaltered under conditions in which water and air have played a prominent part. Some have thought it possible, however, that such conditions may prevail at great depths below our surface so that similar bodies might be formed there. The stony part of the siderolites and aerolites is almost entirely crystalline and in most cases presents a peculiar granular or *chondritic* structure. These chondrules are round or oval granules and are embedded in a matrix of materials similar to them but also frequently containing minute particles of iron and troilite Theories as to how such a condition could arise in their structure are beyond the scope of this book, as indeed are all theories which deal with structure itself. The best we can hope to do is to point out, from astronomical standpoints, from whence such fragments come, not in detail how they were formed.

With regard to the chondritic meteorites, Dr. G. P. Merrill, in a paper entitled *Metamorphism in Meteorites*,[3] concludes partly as follows·

> The chondritic meteorites are regarded as of a tuffaceous origin and their crystalline structure, when such exists, as due to metamorphism in which both heat and pressure have taken part . . . The clear, limpid interstitial glass . . . is shown to have been the last material to congeal . . . . the closing act of the series of changes. The dark glassy interstitial material is considered likewise secondary, a result of partial refusion and alteration. . . The crushing of the individual constituents, while efficacious in the development of a cataclystic structure, is a minor feature and without bearing on the question of the original tuffaceous nature of the stone. The metal is shown to be of secondary origin and its introduction subsequent to the consolidation of the stone in its present form and quite independent of the metamorphism.

He states further that they must have been a result of explosive volcanic activity. But vulcanism is a surface phenomenon, while the meteorites have easily oxidizable materials. Hence they must have been formed in a dry and oxygen free atmosphere This at once rules out our earth unless they were formed when we had no atmosphere. He states that the iron is secondary and owes its reduction to the metallic form probably to the influence of hydrogen. Again.

> The fact that the stones show the effect of heat and rapid cooling may, perhaps, be accounted for on the cometary hypothesis—that their rapid stay in

[3] *Bul. Geol Soc. Am* , **32**, 395, 1921

the proximity of the sun was followed by so rapid a retreat as to prevent a complete recrystallization of the fused material.

L. Fletcher in *Introduction to the Study of Meteorites* makes the following statement as to meteorites in general:

Their general similarity of structure and chemical composition, and more especially the presence of nickeliferous iron in almost every one, suggest that most, if not all of them, have a common source, and that they are chips of a single celestial body.

It is not intended to infer that this was his final opinion as he discusses the other hypotheses and comes to the usual conclusion that perhaps meteorites are parts of comets He quotes work by Wright in 1875 and later by Lockyer which tended to show a similarity between flutings seen in a comet's spectrum and that secured from fragments of meteorites placed in electric glow-tubes. He goes on to say that it may be inferred a comet consists of a swarm of meteorites, not at a high temperature, shining partly by reflected sunlight and partly by the electric glowing of the gases evolved from them by the action of the sun's heat.

As will be brought out fully in the last chapter, while the writer agrees that a comet is actually constituted as just described, at least so far as its nucleus is concerned, he does not believe that this is any answer to the question unless some reasonable hypothesis of how a comet is formed can be given in the first place. We also have positive proof that many meteorites come from interstellar space while all comets seem to belong to our solar system as regular parts thereof.

We have positive information that meteorites and fireballs burst at more than one place along their course.[4] Again we may quote as an example the great fireball of May 11, 1922, which burst for the first time when only 20 per cent of its course had been completed Its height was then about 150 km. It finally burst at only 10 km. above the ground. Its brightness greatly increased after the first explosion. This can be simply interpreted to mean that instead of one large mass from there on very many separate fragments took its place, with enlarged apparent diameter. This fully fits in with our hypothesis. It also proves that enough internal stresses had already been set up to cause such a bursting while at a great height, with corresponding low density of the air.

[4] *Bul. Soc Astr de France*, **38**, 64, 1924.

Coon Butte, Arizona, North Outside View
Photograph by G. K. Gilbert, loaned by Geo. P. Merrill

The effects of the fall of various meteorites upon the surface of the earth have been most dissimilar The velocity which they retain at the end is never very great, but differs for each individual In the Hassle fall stones fell on ice a few inches thick and did not break it nor were they broken. They weighed several pounds each. The 70 pound stone which fell in 1899 at Allegan, Michigan, penetrated sandy soil to a depth of 18 inches and was itself considerably shattered. The largest stone of the Knyahinia fall, striking at an angle of 27° to the vertical, penetrated to a depth of 11 feet. The great meteorite from Cape York, of over 36 tons, was only partially covered, but it lay in a bed of gneissic boulders. It probably fell when there was snow on the ground which broke the effects of its fall. In other words it seems pretty clear that the cosmical velocity of all these bodies has mostly been dissipated by the resistance of our atmosphere and they finally fall with hardly more velocity than if they had been dropped from a considerable height and fallen freely. It also seems fairly certain that unless a meteoric body weighs several pounds on its entrance it will never survive its passage through our atmosphere. We may add that the chances of survival seem to vary, all other things being equal, as the inverse square of the velocity

It is of interest to state that for every meteorite which is found perhaps 100 others have fallen. First, three out of every four fall in the oceans, then all of those which fall in large areas of Asia, Africa, Australia, the central part of South America, the extreme northern part of North America, Greenland, the Antarctic Continent, etc., will have little chance of being found Even in settled and civilized regions, due to forests and sparse population, as well as other factors, most will be lost. The writer has received in the past few years reports of not less than half a dozen, most of which seem certainly to have fallen within a radius of 100 miles, but, due to the mountainous character of the country and the fact that no eye-witness was at the very end, failed to be located. An estimate of one per day would be very conservative for the whole earth.

As for danger, some certainly exists from these projectiles, yet it is so small as to be negligible. In the old Chinese annals an account of a palace being struck was quoted. In the past hundred years there have been several instances of buildings being struck by meteorites, but no proved fatality. In medieval annals we have some accounts of men being killed. Of course if a large meteorite fell upon a house it would be comparable to a solid shot from a large cannon, but as stated the chances are infinitesimally small for its occurring in any given place and year. Accounts differ most inexplicably as to the heat of the masses when first found. Some are said to be extremely hot, but one, the Dhurmsala stone, is said to have been intensely cold when picked up. Doubtless the composition and velocity have much to do with how hot they are when they reach the earth.

No accounts of meteorites would be complete without a description of the remarkable crater found near Canon Diablo, Arizona, frequently alluded to as Coon Butte The following description is mostly abstracted from the work of Dr George P. Merrill, of the United States National Museum,[5] who has made an intensive study of the whole problem, partly on the spot itself.

The crater is a round, saucer-like depression, whose walls rise from 120 to 160 feet above the surrounding plateau, which is about 5000 feet above the sea and semi-arid with an annual rainfall of 8 inches. Within, it is about 600 feet deep, thus being far below the level of the surrounding country. Its diameter is somewhat less than 4000 feet, and its area is about half of one square mile The surrounding region is of limestone and sandstone and without trace of volcanic activity. From borings made 30 miles away it is supposed to be formed of one stratum of limestone and two kinds of sandstone the combined thickness of all being 1600 feet. From general surveys of the region all these strata lie approximately horizontal and are little changed by dynamic or metamorphic agencies The outer rim of the crater is covered by irregular and fragmental blocks which occur in units of every size from hundreds of tons to tiny fragments. Similar fragments are also scattered outside. The larger blocks are all of limestone, for the sandstone has disintegrated more quickly. Vast quantities of crushed sandstone, known as rock flour, is also found upon the rim Trenches and shafts along the rim bring to view masses of both kinds of stone mentioned, mixed without order. On the north rim were uncovered numerous masses of partially oxidized meteoric iron. These occur in such association with the broken rocks as to prove that they were thrown out of the crater at the same time.

As for the interior of the crater, the walls are in general very steep, in many places far too steep to climb. When not covered by frag-

[5] *Smithsonian Misc. Col*, **50,** Part 4, 1908.

ments the original strata show, but in all cases very badly crushed and broken. When one passes beyond the material fallen within from the cliffs, the floor is almost level. But it is easily seen that originally the depression was much deeper and that its present condition has come about by weather action. Large numbers of shafts and drill-holes were sunk in the floor. On an average solid, unbroken rock was encountered at a depth of about 700 feet, showing that the crushing action did not extend further than that level. Meteoric material was encountered in the borings in most cases below about 200 feet and continued as low as 600 feet. It must be understood that only very small fragments could possibly be forced up the drill-holes to the surface.

This is not the place to follow the geological aspects of the crater further except to add that, down to the solid rock levels mentioned above, crushed material was always found, the fragments even when very small being very sharp and angular, as though formed by a sudden blow. What we are here chiefly interested in is, however, the meteoric irons found in and near this remarkable crater. Very few indeed were actually found inside, but how many found outside will now never be accurately known. Dr. Merrill gives an estimate, which he states is little better than a guess, that 20 tons have been found, the number of individuals running into the thousands. The largest one, now in the Field Museum, Chicago, weighs 460 kilograms. The irons have deep concave and convex surfaces, with peculiar holes or pittings. Each iron seems a complete individual, there being no sign of rupture from a larger mass, and no fusion or flow structure from its flight through the atmosphere.

The irons were found scattered over an area of several square miles about the crater, and being picked up usually or at least often by irresponsible persons it is not possible to give a detailed map of their original distribution, even of where the largest fragments were found. This is most unfortunate as we thus lose one of the means of determining from what direction the bodies struck the earth. However an inspection of the crater walls would indicate that, if the crater is due to the impact of a great meteorite or group of meteorites, it came from a direction a little north of west and at a very high angle, perhaps about 70°. This is proved by the worse shattering of the eastern walls and because fragments are thrown to a greater distance on the east than elsewhere. Many of the irons are found actually on the surface, and even if buried originally it was probably to only a slight depth.

What was the origin of this unique crater? The lack of volcanic features and remains in the region, as well as the geology of the crater itself, seem to force us to conclude that it must have been formed by the impact of an immense body from without? In fact there are only two difficulties to fully accepting this hypothesis. First if caused by such an impact why is there not visible proof of the enormous amount of heat that must have been developed? Second what has become of the main mass of the meteorite itself? Indeed the chief reason why so many shafts, etc., were sunk in the floor was the hope of locating a great mass of meteoric iron whose commercial value would be enormous, if of a size at all comparable to that of the crater. Science thus indirectly profited by the prospecting of the engineers, even if the latter failed wholly to locate the mass they were searching for.

In partial reply it may be said that it would not need a mass anything like 4000 feet in diameter falling as assumed to make a hole that large. Yet it certainly would take one a very considerable fraction of that size That it has not been found is not of course absolute proof that parts of it, and large parts, may not lie buried at a good depth below the surface. While there are no volatilization products and little slag so far brought to light, yet some fused quartz has been found.

Merrill's conclusion based upon what is known and upon hypothetical considerations is somewhat as follows If we conceive of a mass about 500 feet in diameter striking with a velocity of 5 miles/second, it might make a crater of the required size As depths below the surface increased, the upward escape of material around the mass would be impeded, and that directly in its path and, to a lesser extent, on either side would become enormously compacted. The heat would produce fusion and partial volatilization, and if moisture were present could account for the pumaceous structure found in the altered sandstone. Moisture being present the impact also would cause steam to be instantly generated, which would rush out with explosive power. This would throw out débris in the manner actually found. It would be to this cause that could be assigned the deposits of rock-flour and fragments in which shale-ball irons are embedded. After the impact a pseudovolcanic condition might remain for a very short time. If this were true a mixture of some of the materials may have flowed

Coon Butte, Arizona, Interior View
Photograph by G. K Gilbert, loaned by Geo. P Merrill

out as a sort of mud. In fact it is possible that all of the meteorites now found outside may have been parts of the original mass and fallen inside, but been thrown out again by secondary action. Perhaps the rest of the main mass was actually volatilized on impact, or there may remain some considerable masses still far below the surface. Finally it may not be necessary to assume such a large original body, as the outrush of gases following the impact of a much smaller one may have been quite sufficient to have blown out a hole as large as actually found. In this case it would not be so difficult to suppose that practically all of the original body was really volatilized and that we might expect to find no large masses below the crater's floor.

In reviewing the evidence it seems that a meteoric origin is the only one that can be assigned, even if there are some real difficulties remaining. The idea has been advanced that the fall of the meteorites took place after the crater was formed, and therefore that their presence around it is purely accidental. Such a hypothesis seems rather improbable. It has further been pointed out that as the large meteoric irons, with one exception, all fell in the same hemisphere with Coon Butte and many of them not very far away, perhaps they accompanied the main mass presumed to have fallen there. Here again this cannot positively be denied, yet the idea while fascinating is most probably not held by men who have specialized in the study of meteorites.

Finally geologists have remarked on the absence of meteorites in the old stratified rocks of the earth's crust. An interesting note on this subject, found in *Nature* **103,** 69, 1919, is abstracted as follows. There it is suggested that the notable absence of meteorites from stratified rock possibly is due to their rapid disintegration. Indeed even when carefully preserved in museums, they often show a deplorable tendency to decay. A piece of meteoric iron was found near Dawson in the Klondike, in the "white channel gravels" which are the oldest high-level gravels of the district and are believed to be of Pliocene age or older. Another similar piece had previously been found not far distant and Canadian geologists are reported to believe that both formed part of the same fall. These pieces were found in 1901 and 1905 respectively. A slice of the mass from near Dawson is now in the British Museum.

CHAPTER XXII

Are Lunar Craters Meteoric in Origin?

There are two theories found in the average work on astronomy to account for the appearance of the so-called lunar craters, the volcanic and the meteoric. We are here concerned only with the latter, and shall try to give the data bearing upon the subject briefly and see what conclusions can be drawn.

Advocates of the meteoric theory[1] suppose that when our system was in process of formation, or at least while there were vast numbers of smaller bodies, from the size of present meteorites to small asteroids, circulating everywhere in most eccentric orbits, that these bodies continually collided with the planets and satellites, adding to their mass. Coming to the time when a more or less plastic or solid surface was formed upon the moon, there still were many such bodies extant, and their fall upon the surface of the moon made the round or elliptical depressions we now call lunar craters. The strongest argument for the meteoric hypothesis is the fact that great difficulties, not to be discussed here, confront the volcanic hypothesis of origin. Positive arguments are: Upon the rims of the craters numerous craterlets or small circular openings are seen Frequently an old crater can be dimly traced under the walls of a newer and higher crater, which by the way is not centrally placed over the older. The craters formed in the earth's surface by the dropping of bombs, when photographed from an aeroplane, strongly resemble lunar craters. Some elliptical craters give the impression of having been formed by a projectile striking the surface at an angle. An example of a similar crater is found at Coon Butte, Arizona. The many streaks leading out radially from certain lunar craters can be explained by the molten material generated by the impact, which would splash out radially.

On the other hand, even if we grant for argument's sake that when the moon's present surface was semi-solid or solid there still were

[1] Consult T. J. J See, *Evolution of the Stellar Systems, Vol.* II, Chap. 14, where the whole question is taken up and a conclusion favorable to the meteoric hypothesis is reached.

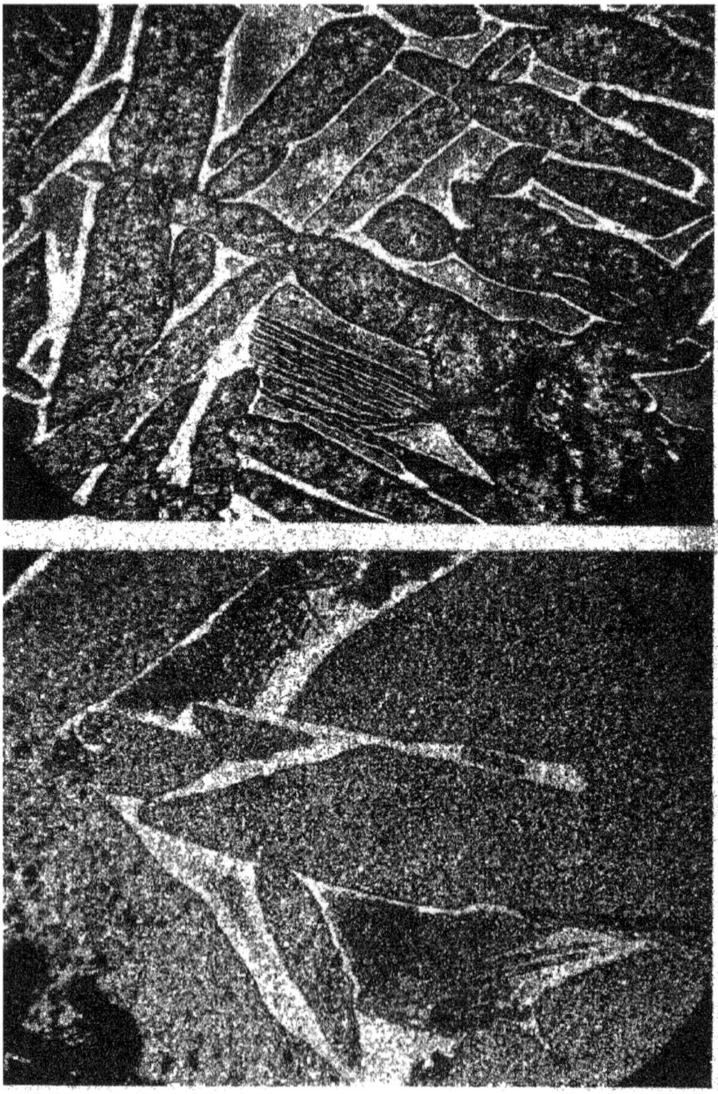

C

D

PHOTOMICROGRAPHS OF METEORITES BY L E JEWELL

Taken by specular reflection From Jorunal of Optical Society of America
C. Canon Diablo Ariz Meteorite *D* Llano del Escado Texas Meteorite

enough such bodies intersecting its orbit to produce the effects seen, there are many weighty arguments against their producing the craters. The distribution is one of the strangest, because some parts of the surface are crowded with craters, others relatively free, and this not including the "seas." If these "seas" then or later contained water we certainly would not expect craters in their bottoms from meteorites. Yet such are found. Then they must have fallen after the seas dried up. But if the seas dried up, it would not have been long, astronomically speaking, before the atmosphere would be lost also. Then there must have been an atmosphere if there were seas. If an atmosphere, it certainly must have produced some retarding effect upon falling projectiles, though small perhaps in the case of any density we can reasonably assume for that of the moon. But if an atmosphere containing water vapor existed, we ought to see some erosion effects upon the awfully sharp and jagged outlines of crater walls. The cliffs and walls are jagged and frequently steeper than we could expect from an impact because if the material was either melted or partly melted, as supposed, due to the blow, it would flow back into elevations with walls less nearly perpendicular, for molten material could not so suddenly solidify as to retain such sharp angles for the sides of the resulting craters. Also central peaks are found in the centers of many of the large craters, with sides having angles upward of 40° to 70°. We do not stress the argument that only one possible case is now known upon the earth because there are sound reasons why this might be the case, no matter how many we see on the moon. Advocates of the meteoric theory advance the argument that the "seas" are areas filled with melted material caused by one or more larger bodies that either buried themselves within the moon or were fused on contact, as indeed they would cause the surface struck to fuse. Admitting this cause and effect, we ask why are the seas lower than the surrounding surface, which in many cases at least appears to be the case? And why in the area near the south pole, where craters occur by the thousands, big and little, are there no seas whatever? As many of these craters are of immense size, why is not this melting action more in evidence? Again it is unreasonable to assume that the fall of very great bodies (i.e., 20 miles in diameter as would probably be necessary for any such action) should be selective both in locality and time. We might have a general distribution in a given epoch, but have no right to assume special distributions in special epochs.

We here desire to call attention to the effects of a minor cause, which absolutely is known still to be going on, but which seems heretofore to have been overlooked. We know that about 20,000,000 visible meteors strike the earth's atmosphere daily. Making allowances for the extra attraction of the earth and the smaller area of the moon, still over one million must strike our satellite every day. Now the earth's atmosphere protects us from the direct striking action of these bodies, for even meteorites have lost most of their velocity before they reach our surface On the moon there is no protection, and there is a continual rain of tiny projectiles moving with velocities up to at least 44 miles per second, admitting only the parabolic limit and nothing higher. These strike the surface at every conceivable angle. Therefore there must be a continual effect, not unlike erosion, going on, because not only would these projectiles break off sharp edges, when striking at an angle, but themselves would be fused and carry off a tiny amount of fused material from the rock struck, if they went through it. In any case there would be a small amount of material melted each time and often distributed. While this process is now so slow as to be almost negligible, one meteor only falling on about every 60 square miles daily, still in ages past there was much more loose material in the solar system unabsorbed than today, as the Planetesimal Hypothesis or any modification thereof would require. Hence such action would then have been greater, due to more material falling per unit of time. The general effect of all this, be it great or small in amount, must be somewhat to smooth the lunar surface. Had it not acted at all that surface would be even rougher than it now is.

CHAPTER XXIII

Origins of Meteors, Fireballs and Meteorites

As Schiaparelli's work has received so much attention his final conclusions in *Sternchnuppen* are here copied word for word:

. . . . The researches and discussions of the chapter have led to no fixed solution of the origin of meteorites, yet they certainly were not without use since they have furnished analogies and contradictions for the different elements of these questions. Summing up the results so far obtained we can set forth the following propositions

1. The grounds which people usually cite against the identity of meteors and meteorites have no compelling force

2 The hyperbolic velocity, which has been observed for certain meteorites destroys every probability of a lunar origin, makes the assumption of a planetary origin for meteorites almost impossible, and does not fit in well with their cometary origin, unless we will assume errors for the observers out of all probability It therefore becomes necessary to place the origin of meteorites in the region of the fixed stars

3. The stellar origin is incompatible with the cometary, for comets belong not to the generality of the bodies forming the solar system, but form in this system a special class for themselves, which has had a common origin with our sun, and in which strongly hyperbolic paths cannot be found

4. The unity of the chemical and mineralogical composition of meteorites would allow the lunar or the planetary hypothesis to appear most probable, if the observed velocities did not furnish a grave contradiction

5. The unity of composition is not incompatible with the derivation of meteors from comets because these have a common origin with our sun.

6. From the assumption of a stellar origin for meteorites it directly follows that they come to us from different regions of the stellar system The mentioned unity of chemical and mineralogical composition one can only explain by this that we assume for the structure of the visible universe a physical and chemical structure like that which meteorites themselves possess.

Newton's conclusions on the subject are taken from an address by him delivered in 1886 and printed in the *Proceedings of the American Association for Advancement of Science*, **35**, 1, 1886. His conclusions as to average heights and velocities of meteors in our atmosphere, their movement around the sun, and their four classical connections with comets will not be repeated. He found, among other things, that ordinary shooting stars do not apparently differ essentially from those in showers; that though meteorites greatly differ from each other in their chemical condition and mineral forms yet they have peculiar common properties that distinguish them from terrestrial rocks, that no organic life in meteorites has ever been found; and that meteors are solid bodies, not little masses of gas. He adds:

> . . . . we may reasonably believe that the bodies that cause the shooting stars, the large fireballs and the stone-producing meteorites, all belong to one class . . . . from the faintest shooting stars to the largest stone-meteor we pass by such small gradations that no clear dividing lines can separate them into classes.

Six good reasons are added for this opinion. He discusses and dismisses the terrestrial, lunar, planetary (including an exploded planet), and solar origin. He concluded in favor of a cometary origin, being careful to say the comet itself could not come from an exploded planet.

Newton accepted provisionally the Nebular Hypothesis as accounting for the origin of comets and then tried to remove the difficulties which arise due to the details of structure of meteorites and the obvious proof they bring us of violent forces at work upon them at different stages of their development. He mentioned an opinion of Daubrée that the union of oxygen and silicon furnishes sufficient heat for making these minerals. And that Reusch argues that the repeated heating and cooling of the comet as it nears and recedes from the sun is enough to account for all the peculiarities of structure of the meteorites. He concluded as follows:

> Suppose then a mass containing silicon, magnesium, iron, nickel, a limited supply of oxygen and small quantities of other elements, all in their primordial or nebulous state (whatever that may be) segregated somewhere in the cold of space. As the materials consolidate or crystallize, the oxygen is appropriated by the silicon and magnesium, and the iron and nickel are deposited in metallic form. Possibly the heat developed may, before it is radiated into space, modify and transform the substance. The final result is a rocky mass (or possibly several adjacent masses) which sooner or later is no doubt cooled down throughout to the temperature of space.

He believes that the comet's approach to the sun is the cause of all succeeding phenomena. His general conclusion is that all meteoric bodies are cometary in origin, as he feels it had been proved for several cases and the analogies were strong for the rest.

Newton two years later followed this up by an important paper[1] in which he sought to determine all he could about the orbits of the meteorites which actually reached the earth's surface. He concluded that nearly all such bodies were moving with direct motion and that their perihelia are not less than 0.5 nor more than 1.0. (If the perihelion of a meteorite's orbit were greater than 1.0 it never could meet the earth.) He discussed the reasons for most stones moving in direct orbits and concludes that the habits of men are not the only reason for this kind being found. So he adds that the direct motion is real, or for some reason the stones moving in retrograde orbits do not reach the ground. His general method of procedure was ingenious, and he concluded that such stones moved in orbits like those of short period comets, not parabolic comets, and hence had low relative velocity.

His argument is greatly weakened in that he used no actual data with regard to observed velocities, nor did he discuss in a thorough manner the chance of survival of a retrograde versus a direct meteorite. As these questions are taken up elsewhere in this book they will not be repeated here. While therefore his work may indeed be considered to prove that nearly all falls of which he could obtain sufficient data were from bodies moving in direct orbits, yet it does not prove in the least that most bodies of like size which enter our atmosphere are moving in direct orbits. As this last is the important question, when origins are being discussed, it cannot be too strongly stressed.

After Newton's death his data were turned over to W H. Pickering and the latter published two papers, which find their natural place for review immediately after Newton's work. In the first paper[2] he discussed the origin of meteorites, mainly from the point of view of monthly distribution. Using Denning's personal data, secured in 1877 to 1889, for his monthly frequencies of meteors, fireballs and slow fireballs, and the *Catalogue of the British Museum* for 1896 for the meteorites, he worked out that the hourly number of meteors and fireballs, when plotted by months, showed similar curves with a strong maximum in the second half of the year. But both slow fireballs and meteorites, though the curves fitted poorly, showed a maximum in May and June. He then tries to show differences between the

[1] *Am Jour. Sci.*, **136**, 1, 1888.

[2] *Pop. Astr*, **17**, 273, 1909

stone and iron meteorites, and quotes the cases of the latter which fell during meteor showers. He concludes that iron meteorites are connected with comets, and that the stony ones were formed of smaller fragments of our earth when the moon was separated from it.

The second paper[3] is a continuation of the first, with reasons given for conclusions stated in the former, as well as further developments. A number of interesting tables and figures are given and the paper should be studied by those interested in the subject. Pickering, probably adopting the views of von Niessl on the subject, which had been published in full three years previously (see page 134 and 137), correctly pointed out why we should expect to find few meteorites moving in retrograde orbits that had actually reached the ground as such. His argument continues as follows:

> If all meteorites before their fall were moving in direct orbits, even those whose fall is observed between noon and midnight . . . . must overtake the Earth. . . . . On the other hand those observed between midnight and noon must be overtaken by the Earth, and be moving more slowly than it. Any variation from this rule implies a highly inclined or very eccentric orbit

Tables II and III are from Newton's (corrected) data and give much information about the probable orbits of the meteorites in question. This statement was made, however: "Comparatively few quits, i.e., points from which the meteors come, lie far north or south of the ecliptic." This point being very important in his theory, as small latitudes for q necessitate small inclinations, the values of q were counted in groups, 107 in all being available. These were from 90° to 71° 6 meteorites, from 70° to 51° 9, from 50° to 31° 14, from 30° to 10° 38, and < 10° 40 meteorites, total 107 with a mean q = 23.7°. While the values of q are not the inclinations themselves, still they are a function of the latter. It is therefore submitted that there is little warrant for the statement that "few quits lie far north or south of the ecliptic", as 15 lie more than 50° away, and 29 lie over 30° away, out of only 107. There are also 29 quits south of the ecliptic only four of which are 30° or over. This throws the other 25 values near the ecliptic, a mere result of most of the available observations having been made in the northern hemisphere, and these 25 help to make the fairly low value found for q, i.e., 23 7°.

[3] *Pop. Astr*, **18**, 262, 1910

The rest of the article in question is taken up with the results based upon the table II and certain other assumptions from all of which is drawn the conclusion that the stony meteorites move on an average in an orbit with the elements:

$$\alpha = 1,\ \pi = 250°,\ \Omega = 115°,\ i = 6°,\ e = 0.14 -$$

He considered this a strong support to his hypothesis that the stony meteorites were formed when the moon burst away from the earth, but that the iron meteorites are probably from comets. He ended the article by a word about the relative numbers of meteors we should meet in the morning compared to the evening, which he deduces should be 6:1 on the assumption of parabolic velocity. He then stated the observed ratio is only 2:1, hence there are three meteors moving in direct to one in retrograde orbits.

This conclusion is partly erroneous for any given place on the earth because it was long since proved by Newton, Schiaparelli, von Niessl, and others that the ratio of the number of meteors seen at any hour was a function of the zenith-distance of the meteoric apex as well as other factors and not only the velocity alone These older observers deduced equations, all based upon the assumption of meteors moving equally in all directions which presupposes as many direct as retrograde. Indeed in 1922 Hoffmeister extended this work in a noteworthy way, but no one besides Pickering has deduced any such preponderance of direct meteors over retrograde, no matter what has been their method of investigation. In a more recent paper[4] he greatly modifies this opinion seeking to prove from some of the published results of radiants by Denning and orbits of meteor streams by Olivier that observations can be made to show an equality between the numbers of direct and retrograde meteors. The writer does not dissent from this result but admits inability to follow the argument leading to it, particularly as relating to his own orbits. In this last paper Pickering still holds strongly to his hypothesis of the origin of stone meteorites, with the low velocity it implies.

Turning now to the work of von Niessl,[5] which was indeed all published considerably before any of that by Pickering just mentioned, we have seen in Chapter XII that the former showed most clearly how

[4] *Pop Astr.*, **27**,, 203, 1919

[5] *Smithsonian Misc. Col*, **66**, 16, 1917.

the depth of penetration of meteors, fireballs and meteorites varied according to their velocities, as well as their masses. Hence we understand why even if there were a Leonid and a Bielid, for instance, both equally large and the latter could reach the ground as a meteorite yet the former could not. He states that the heliocentric velocity, resulting from the observed velocity, in most cases far exceeds the parabolic. Also that this is the case for those orbits which can be most accurately derived and for which the velocity stands the best chance for good determinations, namely those from the anti-apex. For 26 of the best determined cases this comes out 59.05 km./sec. And for those cases with enough material on which to base a sound conclusion the hyperbolic orbit is a certainty. For 154 large meteors whose paths are derived from the older material or were first computed by von Niessl himself the heliocentric velocity came out 59.8 km./sec. His conclusions are now given in his own words:

Therefore, in general, the large meteors are undoubtedly of interstellar origin As opposed to this conclusion, we find that streams of shooting stars pursue the same orbits as certain well known comets of well known periodicity. They are, therefore, interplanetary shooting stars Hence we are inclined to consider the large meteors as interstellar, but the smaller shooting stars as interplanetary Still we must call attention to the undeniable fact that most of the radiation points of meteors and detonating fireballs as well as the other large meteors, as far as they can be safely determined, agree with well established shooting star radiants. It is difficult in such cases to ascribe interplanetary orbits to the corresponding small phenomena when the hyperbola described by the large meteors issues from the same radiant points in the cosmic space.

In attempting to solve this apparent contradiction, one might perhaps assume that we have included under the name of shooting stars different phenomena that are only superficially similar, but whose dynamic base and cosmic significance are probably not all similar. From the many experiences of the last ten years we may at least draw the conclusion that in the phenomena as a whole both the limited and minor interplanetary meteors make themselves felt as well as the extended interstellar meteors having particles of the greatest variety as to size, mass, quality, and velocity.

One more older quotation[6] from Schiaparelli will be given as to meteorites:

Whenever we have been able to investigate with any approximation the velocity with which a meteorite or a group of meteorites have penetrated into

[6] *Sternschnuppen*, 207

Region of Milky Way South of Alpha Cephei, Showing Two Dark Objects, One That in Plate 3b

Exposure $6^h\ 2^m$ by E E Barnard Taken October 1, 1910, at Yerkes Observatory

the atmosphere, we have always found that the corresponding absolute velocity is greater than the parabolic would be.

Schiaparelli again in 1910,[7] discussing the question, strongly confirms the view that meteorites come from interstellar space, and hence with hyperbolic velocity. He goes on to say that experiments at Potsdam by Scheiner proves that the two kinds of gases which usually are found in the spectra of comets (hydrocarbons and oxides of carbon) are equally contained in meteorites in a state of occlusion. Vogel by superposing the spectra of the two gases mentioned and putting a fragment of a meteorite in a Geissler tube obtained identical spectra. Therefore, Schiaparelli continues, comets and meteorites may be bodies of identical or similar nature, although in appearance differing in their manifestations. Then:

The meteorites may be comets of other suns, which, under their heating action have already by frequent and great emissions of jets and tails lost all or nearly all their contained gases, while our sun has not yet extracted from all its comets and dispersed in space the totality of the gas that they originally contained. Finally, comets and meteors may differ among themselves only in the diversity of the phases attained in their evolution.

He gives as a possible hypothesis that comets form with the sun a kind of stellar current (like Kapteyn's currents), while the stone meteorites and the iron meteorites may in space have formed part of at least two other such currents.

The latter part of this is wholly unprovable in the present state of our knowledge, and will not be discussed. However, there is another point brought out in this article about the short period comets which has a direct bearing upon the conclusions of Newton and W. H. Pickering just discussed. It has to do with why short period comets all have direct motion and rather small inclinations. He says that as to the predominant frequency of small inclinations with the short period comets we should not say that the inclinations are small because the comet is periodic, but that the comet has become periodic (i e., with short period) because its inclination was small before it met the perturbing planet and remained small afterwards. He adds that for comets with retrograde motion the same perturbations would not so shorten the period, hence we have none of excessively

[7] *Bul. Soc. Astr. de France*, **27**, 250, 1910.

short periods, even though we have periodic comets as Temple's and Halley's both with $\iota = 162°$. Their periods are respectively 33 and 77 years.

This has a bearing upon the distribution of meteorites. For if, for arguments sake, we agree with W. H. Pickering to the extent that most meteorites meet us with direct motion and low velocity, then if some of the meteorites once formed parts of comets belonging to the solar system, those moving in direct orbits of small inclination would have most chance of being turned into short period orbits, and hence would pass the earth oftener in any given time. This would give us far more chance of picking up this kind than those moving with long period retrograde motion. But even with both these admissions—which we are not indeed prepared to agree to—it can thus be shown that the present preponderance of direct motion in meteorites does not necessarily confirm an original preponderance of numbers, nor an actual present majority, if we could count all now in our system which originated there. This is an additional argument to the most weighty of all, namely, that those of highest velocity have little chance of reaching our surface.

Finally we mention again the work of Hoffmeister (see Chapter XVI) by which he is led to assign a highly hyperbolic velocity to the average meteor observed during the course of his work. If his conclusion is confirmed definitely by future researches the origin of many or most meteors, as well as meteorites, must be assigned to interstellar space—or much more correctly speaking to other systems than our own. It is largely due to this piece of work that the final opinions herein stated differ from those of Eberhard in the last edition of Newcomb—Englemann's *Popular Astronomy*. This was due to Eberhard not having Hoffmeister's results when the chapter was written. Occasion is here taken to refer all readers to the excellent chapter on Comets and Meteors, contained in the above work. They furnish a brief but most complete survey of the subject. Indeed the writer's labor in preparing this book would have been considerably lightened had he been able, before this was ninety per cent completed, to obtain a copy of the astronomy just mentioned.

Brédikhine argued against the reality of observed hyperbolic velocities for meteorites and fireballs, but by a curious slip his figures prove exactly the opposite to his conclusion and, if his argument is sound, then the velocities are even more hyperbolic. This slip was pointed out by Hoffmeister.[8]

Nearly every astronomer who has worked on meteors, and many who have not, has expressed opinions on the subjects treated in this chapter. Many of these opinions are based on partial data, some on mere guesses—indeed their number is so great that it is absolutely impossible to give a new one. The sun, moon, earth, planets, comets, stars, etc., have all been given as the birth-place of meteors, until there is no body left in the universe to which a writer could today refer their origin and claim originality! And it is almost needless to add that most of these opinions contradict one another. An attempt has therefore been made, in this chapter, to review those which appear most worthy of attention and advanced by men who have studied the question.

[8] *Loc. cit.* E 2.

CHAPTER XXIV

CONNECTIONS BETWEEN METEORS, FIREBALLS AND METEORITES CONCLUSION

In the course of this book descriptions of all these classes of bodies have been given, along with many theories of their origins and possible connections. Also their physical characteristics have been described. It is therefore our final task to attempt to show whether these bodies are all of the same general type, merely differing in size, or whether there are greater differences even extending to a wholly different origin for each class or for individuals of each class. Let us first sum up a few facts that are denied by no one, for when we came to the theories most violent differences of opinion are found.

1 Many fireballs have appeared during several meteoric showers, coming from exactly the same radiant as the smaller bodies.

2. During considerable showers only one single meteorite has been known to fall (on November 27, 1885). Unfortunately we have no data whatsoever as to its path or appearance before it struck the ground. It is true, as stated before, there have been several coincidences in dates of the fall of meteorites with the time when the earth was passing the nodes of other showers, but no definite proof of connection in any case.

3. In apparent magnitude we find bodies varying from the faintest visible in the telescope up to fireballs and meteorites which to observers appear as large as the full moon, with every possible gradation in between.

4. The monthly numbers of meteorites show a strong maximum in May and June, which is quite contrary to the monthly numbers of meteors as shown in Chapter XVI. The monthly distribution of fireballs follows that of meteors. For comparison purposes, however, there are only available somewhat less than 400 meteorites, scattered over a century and a quarter.

5. Meteorites are most frequent during the period from noon to midnight, namely, when bodies must overtake the earth in order to be seen at all. Hence they move with much less average geocentric velocity than those bodies which meet us from midnight to noon,

when we are on the forward side of the earth. Therefore assuming two meteorites of equal mass, moving in parabolic orbits, one coming from the apex, one from the anti-apex and both with parabolic heliocentric velocity, the geocentric velocities will be about 72 km./sec., and 16 km./sec., respectively, when the earth's attraction is allowed for. But kinetic energy is proportional to velocity squared times mass or v^2m. In this case v_1^2 is 20 times as great as v_2^2, and this represents the energy developed in the two cases by striking into our atmosphere. We may probably assume that the chances of destruction are therefore about 20 to 1 in the two cases or directly as the squares of the velocities. This seems to be the only reason, unless we postulate a very different chemical constitution, why of all the fireballs seen in the great Leonid showers not one has ever reached the ground.[1] Yet for the slow moving Bielids there would be a good chance for a fall, as perhaps was the case with the Mazapil meteorite already referred to. However, we would not expect a fall from the η Aquarids, Orionids, or even the Perseids, as all these have radiants rather near the apex. Incidentally fewer fireballs are furnished by these than by the Leonids, but this may be purely due to the vastly greater number of the latter seen in showers. The Geminids, however, might furnish meteorites, and there are several falls early in December which might as well belong to that group as not We simply have no data to prove or disprove it.

6. We are further led to believe that the differences are merely due to size because as a class meteors disappear much higher than fireballs; fireballs of course higher than meteorites, which latter actually reach the ground.

7. It has been proved that for several prominent meteor streams the individuals follow approximately the same orbit as some well-known periodic comet.

8. We have proof that several comets have divided their nuclei into one or more parts actually before our eyes; that in one case the two new parts appeared again as small separate comets somewhat further apart than on the previous appearance; that certain periodic comets have wholly disappeared with no possible reason for us to give except dissolution.

[1] The writer claims no credit for being the first to bring this out, it having been advanced formerly by others without any law being actually given, so far as known, yet the idea has not received proper recognition

9. All evidence points to the fact that very many fireballs come to us in hyperbolic orbits, and must hence be formed without the solar system. This includes some meteorites also.

10. Most astronomers, who have worked in this branch of the science, would unhesitatingly add that the radiant points of fireballs and meteorites often are connected with those of meteors. The writer will neither affirm nor deny this as a general proposition. Nevertheless he is certain that a large part of the so-called coincidences occur with meteor radiants whose existences are not proved in a scientific manner. Until we have better catalogues of radiants of showers to use for comparison, this point must be given very little weight.

Coming now to the results of certain investigations, which while most instructive and worthy of study we cannot accept as fully proved because certain (if reasonable) assumptions were made, we find that the theoretical work of Newton at one time tended toward the hyperbolic velocity for meteors, while that of von Niessl both for fireballs and meteors tended in the same direction. Finally Hoffmeister has found from his own splendid observations, recently made, and also from a study of the old series of Coulvier-Gravier and Schmidt, that the phenomena of daily and yearly variations are best explained by assuming a velocity of about 2.4, which is very strongly hyperbolic.[2]

Finally with regard to the constitution of meteorites, as has been shown, many if not all once must have formed part of some larger body in which temperature, pressure, crystallization and other physical and chemical processes took place. Such being the case, is the nucleus of a comet the place to look for such actions? Certainly not, if as generally considered it is made up of numerous small solid bodies surrounded by tenuous gases. If on the contrary the nucleus is made up of bodies of the order of small asteroids, then we might expect such effects to take place and their results to be shown in the broken fragments. But we have little belief and no proof that the average nucleus does consist of bodies so large.

This point of view is urged because it is believed to be utterly unscientific to conceive of space containing innumerable very small

[2] In *Mitteilungen der Sternwarte zu Sonneberg*, *Nr.* **5**, 1924, Hoffmeister publishes 58 orbits of meteors, based upon corresponding observations, many of which proved to have hyperbolic velocity.

MESSIER 8, CONSISTING OF THE NEBULA N G C 6523 AND THE STAR CLUSTER N.G.C. 6530, BY J. C. DUNCAN AT MT WILSON OBSERVATORY

solid fragments which were formed directly from tiny separate masses of gas. In other words if solid masses of small size exist they come from broken parts of larger masses where enough gas had collected to gradually condense into first a liquid and then a solid body.[3] But the writer believes absolutely that there is a minimum mass required, which is vastly greater than the size of any meteorite ever found on our earth. If we follow this reasoning to its apparent conclusion we decide that comets either contain much larger solid masses in their nuclei than usually is admitted, or that comets themselves were formed in a catastrophic manner from larger bodies.

The writer firmly believes, as do most others, that the simplest form of cosmical matter is gaseous and that it is illogical to begin any hypothesis with larger units of solid matter, unless we have some reasonable explanation of how the solid units came from the gas. He is thus absolutely unwilling to say that meteorites and meteors come from comets, without at the same time admitting we are only pushing back the question of their origin one step and realizing that we must explain the origin of solid nuclei in comets if we wish to solve the problem. It seems this point of view has been quite generally ignored by astronomers.

We know, however, that asteroids, which on many grounds are believed to be solid bodies, vary in size from Ceres, whose diameter is nearly 500 miles, down to the faintest now discoverable, which we compute may be 5 or 10 miles in diameter only. These bodies have orbits of various inclinations and eccentricities, and we might add a link to our evidence if some asteroids were found whose orbits were quite cometary in character. For instance we find asteroid No. 944 having $\iota = 43°$, $e = 0.66$, $a = 5.6$ and $q = 1.9$, an orbit which certainly has every affinity with that of a periodic comet. Again asteroid No. 945 has $\iota = 33°$, $e = 0.16$, $a = 2.6$ and $q = 2.2$, an orbit which has a very considerable inclination. Asteroids No. 719 and 887 have eccentricities of 0.54 and 0.53 respectively. Eros also has a peculiar orbit, not unlike a comet's in some respects, and we might

[3] In confirmation of this view we refer to a paper on the early history of the solar system (*Monthly Notices R A.S* 78, 424, 1918), by Harold Jeffreys, a distinguished English mathematical astronomer, who concluded that no planet or satellite, whose present diameter is less than 1000 km., could ever have been gaseous, as its gravitative power would have been unable to hold it together.

quote several astroids whose orbits have large values for ι or e The asteroids mentioned above are faint and must be bodies of small size. There is absolutely no doubt that if No. 944 had had a coma when found it would have been called a comet by everyone. Is it possible that it is an example of a comet with only the nucleus left? However this question may be answered, the possible connections of comets and asteroids have been made slightly more probable by the discovery of this asteroid. Such a proved connection would have the tremendous advantage of allowing us to see how a comet could have a nucleus, made up of one or many solid parts. Though even here it is doubtful if we succeed in entirely getting away from a catastrophic origin.

There recently appeared[4] an observation by Comas Sola of Barcelona in which he stated that Asteroid No. 224—Oceana—on December 13, 1923, was photographed with a faint halo around it, whose diameter was 30″. At that date Oceana was 167 million miles from the earth and of magnitude 11.6, hence the halo was 24,000 miles in diameter. The planet belongs to the middle group of asteroids. This group has a high percentage of large bodies and contains most asteroids with very eccentric orbits. Oceana itself, however, has an inclination of only 6°, $e = 0.04$, and $P = 4.3$ years. It is stated by M. Thiele that an accurate study of its motion, as yet only partially investigated, might disclose changes in some of its elements if the halo indicates a disintegration of the little planet. This observation, thought to be unique, is mentioned at length because if confirmed another link of evidence would be available for possible connection between comets and some asteroids. It is to be remembered, however, that all asteroids so far discovered have direct motion, while comets are about equally divided between direct and retrograde. Some of the smaller, outer satellites of Jupiter and Saturn, and all of the satellites of Uranus and the one of Neptune move in a retrograde direction. The eccentricity of the eighth satellite of Jupiter is 0.38, while two of the others mentioned have $e = 0.16$ and $e = 0.17$ respectively. As these last are justly considered permanent members of the solar system it proves at least that the differences between the orbits of comets on the one hand and the orbits of some asteroids and satellites on the other are not so great as used to be supposed.

[4] *Pub. Astr. Soc. Pac.*, **36**, 88, 1924.

Leaving the question of ultimate origin unanswered we are bound to admit that in certain proved cases streams of meteors, with fireballs mixed in, do follow orbits practically the same as those of the comets which usually are called their generators. Hence to periodic meteor streams with comets attached we must assign an origin within our system because we have elsewhere shown the proof by Stromgren and others that comets belong to the solar system. This connects comets, fireballs and meteors in a bond we cannot argue away.

But if Hoffmeister is correct in saying that the average velocity of all meteors (except those of a few periodic streams) is about 2.4 or 71 km /sec., and we know there are many of those observed by him whose orbits are elliptical besides the three streams excepted from his lists, then there are many with velocities much greater than 2.4 to make up for the elliptical ones with velocity less than 1.4, i.e., the parabolic velocity.

A test is here suggested that is believed to be original. If a group of meteors furnish a good radiant one year, but no trace of this is reliably found in very many succeeding years, on an average this might be suspected of being a stellar current. But if the radiant is active a few years afterwards, but with a quiescent period between, then it is almost surely a periodic phenomenon and belongs to the elliptical class, and is a member of our system. If however it appears year by year, for any specific case it may be either stellar or interplanetary in origin. But for a great number of such cases the presumption would be that most are interplanetary because it would be an unwarrentable assumption that space is filled with a very great number of currents, which would by chance be intersected by that part of our solar system containing our orbit, and which would all be so wide that it would take us many years to pass through and still be densely enough filled with meteors for us to meet them yearly in sufficient numbers to give a radiant point. This does not in the least deny the existence of vast numbers of stellar currents but no valid reason is seen for believing that their average width should be so great that it would take our system 10 or 100 or some longer period of years to pass through. It is therefore believed that this test, intelligently applied to scientifically observed radiants, will in the end throw much light upon this question.

For isolated streams which furnish enough bright meteors to give a chance for success it is almost redundant to say that observations of their velocity, made by means of the photographic method, will eventually give us direct information on the subject. Everyone may eagerly look forward to the obtaining and publication of authentic data of this kind As has been explained the mental estimates of even the best observers are too rough to be of great value, especially for the average meteor whose duration is of the order of one half second only. For slow meteors and fireballs, with long durations, such estimates become more valuable.

Some astronomers, though it is thought relatively few, have believed meteorites to be the product of lunar volcanoes, and one or more have believed that they were formed when the moon burst away from the earth after the latter had formed a crust. The number of astronomers who believe in the actuality of the latter event is so few that it is scarcely necessary to discuss it. But it may be said that if such an event did take place the orbit of every meteorite so formed would have to cut our orbit at the point where the explosion took place.[5] Also relative velocities of large amounts are impossible. The holders of this hypothesis, besides believing in a most improbable event, wholly ignore the observations made on very many fireballs and some meteorites, which prove immensely greater velocities than all the planetary perturbations combined could ever give to an original fragment of our earth.

As for the lunar origin of meteorites, it is equally impossible to use it as a general explanation of their existence. Even if the formations on the moon are true volcanic craters and ever had had explosive energy enough to drive material away from the moon—both incidentally questionable hypotheses, particularly the second—the material would again travel with relative velocities far too small to fit our observations

To assign them to volcanoes on planets is first assuming there are such volcanoes, of which we have no proof, and if the major planets are still plastic a pretty certain disproof so far as they, otherwise the most probable generators, are concerned.

The only place that eruptions of sufficient violence are known to exist is upon the sun This statement excludes comets which have already been fully discussed Here the eruptive prominences might indeed be conceived, when of extraordinary violence, to project matter out which might never return, i e., move with hyperbolic velocity.

[5] Lunar and planetary perturbations would prevent this statement from being exactly fulfilled, after great intervals of time had elapsed.

Yet this velocity must be greater than 384 miles / second at the sun. However, under no conceivable circumstances could the intensely heated gaseous material of a solar prominence suddenly be converted, when cooling in empty space, into a rock with complicated structure or into an iron meteorite showing crystalline structure. In fact it is far more likely that the gas, released from external pressure and with a violent projective force impelling it, would simply dissipate into separate molecules rather than that it would condense into a solid or a group of solids of small size.

We are then thrown back upon the cometary origin of meteorites or an origin enough similar thereto which would produce solid masses of complicated structure. Yet it has been frankly said that this is only one step—we then have to explain the comet's origin. The above refers to such meteorites as move with hyperbolic velocities.

For the large number which belong to this latter class we are forced to conclude that the same types of bodies are built up in similar ways, in other stellar systems. We have already seen how planetary perturbations can force a meteor in an elliptical orbit to acquire hyperbolic velocity and leave the solar system. Of course this will eventually enter some other system. In other systems giant planets could impress greater velocities than even could Jupiter, so that meteorites should escape from such systems with great hyperbolic velocities is reasonable enough to suppose.

Returning now to the general proposition of the origin of comets, meteors and perhaps the smaller asteroids if not all of them, the writer sees no possible explanation on the basis of the Nebular Hypothesis, unless we can assume it is not an infrequent thing for a completed or an almost completed planet to explode.

An objection will here be raised that we do not find examples of hyperbolic comets. This can be partly answered in several ways. First that comets are far less frequent than meteorites and fireballs, when we remember that at least half the visible comets of any size are discovered, whereas only one meteorite in every 100 at most is found. Secondly that comets are a far more unique or perhaps specialized form of creation than are meteorites, and that it is conceivable that meteorites could readily be formed in some system without it being possible for comets to be. Thirdly that it is not impossible at all that sooner or later an hyperbolic comet may be discovered, for accurate and authentic observations do not run back more than two centuries or more. And lastly it might be hazarded that while simple units like a meteor or meteorite can move unimpeded from system to system there may be some unknown causes which prevent complex bodies like comets from making the journey.

We are then brought back to a catastrophic hypothesis. Of these latter, that by Moulton and Chamberlin, the Planetesimal Hypothesis, seems best to explain the meteoric phenomena exhibited by all classes of bodies in our system. But we hasten to add that if we accept it here we must accept it in many other stellar systems, the solar system ceases to be unique as many recent writers prefer to consider it. It seems, however, that the great frequency of novae indisputably prove, despite all the calculations based on the assumed number of suns and the theory of probability, that catastrophics are actually numerous enough in our universe to give every star a chance for such an event sooner or later. The writer has no intention of attempting to discuss the apparently sound reasoning on which many eminent writers have calculated the small changes for such a catastrophe in a given case, but does reiterate that every year we actually see novae and that so far as the arguments presented in this book are correct the catastrophic origin seems the reasonable one. Again this in nothing denies that a star or a double star is formed by the condensation from a nebula as in the Nebular Hypothesis, but only affirms that once such a star is formed for it to generate a system containing all the component bodies we find in the solar system, for instance, the Planetesimal Hypothesis is the one to which we are logically led when we attempt to explain meteoric phenomena.

To sum up briefly, it seems probable:

1. That there is no difference except mass and geocentric velocity between meteorites, fireballs, and meteors.

2. That representatives of all these three classes of bodies seem intimately connected with comets in our solar system. And that possible connections between small asteroids, satellites, and comets' nuclei are appearing, in view of recent observations.

3. That large numbers of meteorites, fireballs, and meteors also come to us from outer space. This infers conditions in numerous stellar systems enough similar to our own to generate similar bodies.

4. That the Planetesimal Hypothesis seems the most probable one in the light of all meteoric phenomena. And further that this, or some similar catastrophic, origin must be assigned to a large per cent of all stellar systems, if we would explain the number of bodies we meet which come from outer space.

INDEX

Sans Tache

Sans Tache

IN THE "elder days of art" each artist or craftsman enjoyed the privilege of independent creation. He carried through a process of manufacture from beginning to end. The scribe of the days before the printing press was such a craftsman. So was the printer in the days before the machine process. He stood or fell, as a craftsman, by the merit or demerit of his finished product.

Modern machine production has added much to the worker's productivity and to his material welfare; but it has deprived him of the old creative distinctiveness. His work is merged in the work of the team, and lost sight of as something representing him and his personality.

Many hands and minds contribute to the manufacture of a book, in this day of specialization. There are seven distinct major processes in the making of a book: The type must first be set; by the monotype method, there are two processes, the "keyboarding" of the MS and the casting of the type from the perforated paper rolls thus produced Formulas and other intricate work must be hand-set; then the whole brought together ("composed") in its true order, made into pages and forms. The results must be checked by proof reading at each stage. Then comes the "make-ready" and press-run and finally the binding into volumes.

All of these processes, except that of binding into cloth or leather covers, are carried on under our roof.

The motto of The Williams & Wilkins Company is *Sans Tache*. Our ideal is to publish books "*without blemish*"—worthy books, worthily printed, with worthy typography—books to which we shall be proud to attach our imprint, made by craftsmen who are willing to accept open responsibility for their work, and who are entitled to credit for creditable performance.

The printing craftsman of today is quite as much a craftsman as his predecessor. There is quite as much discrimination between poor work and good. We are of the opinion that the individuality of the worker should not be wholly lost. The members of our staff who have contributed their skill of hand and brain to this volume are:

Composing Room: Henry Shea, Edward Rice, Walter Phillips, Ferd Schilling, George Moss, George Behr, William Fite, Herbert Leitch, Roland Stultz, John Flanagan, James Armiger, Ray Kauffman, Zeddie Breithaupt, Andrew Rassa, Harry Harmeyer, Kurt Woernle, William Koch, Ernest Salgado, John Dotterweich, Richard King, Harry LaMotte, Theodore Nilson.

Proof Room: Sarah Katzin, Alice Reuter, Mary Reed, Lucille Bull, Arthur Baker, Ruth Treischman, Ethel Strasinger, Angeline Eifert, Audrey Tanner, Edna Clarke, Lillian Gilland, Dorothy Strasinger,

Press: Leo Ledlich, Raymond Bauer, Edward Smith, George Lyons, Henry Eckert, Robert Gallagher.

Cutter: John Murphy.

Keyboard: Harry Susemihl, Minnie Foard, Anna Kelley, Hannah Scott, Eleanor Luecke, Catherine Kocent.

Casters: Kenneth Brown, Ernest Wann, Mahlon Robinson, Charles Aher, George Smith, Frank Malanosky, Martin Griffen, Henry Lee.

Folder: Lawrence Krug.

Milton Keynes UK
Ingram Content Group UK Ltd.
UKHW012216020224
437193UK00004B/135